Quantum Mechanics of the Diatomic Molecule with Applications

Quantum Mechanics of the Diatomic Molecule with Applications

Christian G Parigger and James O Hornkohl
University of Tennessee Space Institute, Tullahoma, TN, USA

IOP Publishing, Bristol, UK

ISBN 978-0-7503-1989-8 (ebook)
ISBN 978-0-7503-1890-7 (print)
ISBN 978-0-7503-1941-6 (mobi)

DOI 10.1088/978-0-7503-1989-8

Version: 20191101

IOP ebooks

British Library Cataloguing-in-Publication Data: A catalogue record for this book is available from the British Library.

Published by IOP Publishing, wholly owned by The Institute of Physics, London

IOP Publishing, Temple Circus, Temple Way, Bristol, BS1 6HG, UK

US Office: IOP Publishing, Inc., 190 North Independence Mall West, Suite 601, Philadelphia, PA 19106, USA

To John, Anna and Melissa

To Justin, Jason and Jola

Contents

Preface

The book notes from J O Hornkohl and extensive scientific discussions and research engagements in my work at the University of Tennessee Space Institute, Center for Laser Applications, motivate completion of this ebook. Communication exchanges occurred since the spring of 1987 and continued regularly until winter 2017 [1]. Over the years, several colleagues and postgraduate MSc and PhD students contributed to applications of fundamental insights in the physics of the diatomic molecule. Thanks go to David Plemmons, Guoming Guan, Ying-Ling Chen, Wenhong Qin, Ivan Dors, Alexander Woods, David Surmick, Michael Witte, Ghaneshwar Gautam, and Christopher Helstern.

Significant emphasis has been made on the application of the diatomic spectroscopy predictions in analysis of experimental data. For this reason, this ebook includes several chapters on applications in studies of diatomic molecules, especially important molecules such as cyanide (CN), aluminium monoxide (AlO), diatomic carbon (C_2), titanium monoxide (TiO), and hydroxyl (OH), but also selected work on other diatomic molecules.

The text introduces insights that are essential in utilizing the inherent symmetries associated with diatomic molecules. Consequently, line positions and strengths associated with transitions from lower and upper state-manifolds are determined without invoking approximations that separate vibrations and rotations of diatomic nuclei from electron motion based on mass. The approach utilized in this work makes use of the separation of angular coordinates from electronic vibrational coordinates. Consequently, the volley of selection rules for diatomic spectroscopy is no longer required, including methodologies that rely on so-called reversed angular momentum techniques.

The work summarizes well over 30 years of quantitative analysis of temporally and spatially resolved experimental records, almost all of the experiments discussed in this ebook were conducted at the Center for Laser Applications (CLA) at the University of Tennessee Space Institute (UTSI). Applications include understanding on non-equilibrium fluid and plasma physics and interpretation of stellar astrophysics spectra. In several cases of laser-induced plasma investigations, both atomic and molecular signatures or superposed spectral characteristics from molecules and atoms can be identified. Analysis of such superposition spectra requires accurate knowledge of wavelength positions and transition strengths. The revival and replacement of electrical-spark spectroscopy with laser-spark or laser-plasma spectroscopy for quantitative elemental composition analysis since the mid-1990s, viz. laser-induced breakdown spectroscopy (LIBS), extends into increased interests in molecular laser-induced breakdown spectroscopy since give-or-take the mid-2010s. From an analytical and practical point of view, the requirements can be reduced to availability of a set of diatomic line-strengths in tabular form along with programs that are designed to appropriately read the records. However, this ebook provides a reasonable account of the quantum mechanics of the diatomic molecule along with selected applications that were important for motivating a consistent approach and for analyzing recorded data sets from various experiments in the CLA laboratories.

The challenge to the work has been the prediction of spectra with focus on diatomic spectroscopy. The aim of the lifetime work of Jim Hornkohl is the design of an algorithm to predict and fit computed and measured molecular spectra to provide inferences on parameters such as excitation temperature. The means to accomplish goals for various diatomic molecules are the consistent application of standard quantum theory of angular momentum. During his career, Jim engaged in efforts to overcome techniques such as Van Vleck's reversed angular momentum approach based on angular momentum commutators. The apparent difficulties included the battles with the established practice to predict and compute spectra and design programs despite the mathematical inconsistencies associated with the reversed angular momentum practice. The experimental investigations, and again the stimulating discussions, motivated refinements such as enlarging the data sets for the CN, C_2, or TiO diatomic molecules. In turn, the discussed applications in this book are intended to alleviate analysis of diatomic spectra composed of superpositions of a significant amount of transition lines within typical resolution for laser-plasma emission spectroscopy, to name but one example.

Christian Parigger
August 2019

Reference

[1] Parigger C G and Nemes L 2017 *Int. J. Mol. Theor. Phys.* **1** 00105

Acknowledgments

CGP and JOH appreciate the support by the Center for Laser Applications (CLA) at the University of Tennessee Space Institute during over 30 years of research engagement in spectroscopy. The research led to this book that describes how quantum mechanics can be used to predict diatomic molecule spectra in a gaseous state. The book provides a comprehensive overview on diatomic molecule fundamentals and emphasizes the applications of spectroscopy predictions in analysis of experimental data recorded at the CLA. Both CGP and JOH are happy to report interests and contributions by postgraduate students. Moreover, CGP thanks researchers and colleagues at the CLA and at various international conferences for extensive discussions.

Author biographies

Christian G Parigger

Photo courtesy of UTSI photo albums: L Horton

The research interests of Dr Christian Parigger include fundamental and applied spectroscopy, nonlinear optics, quantum optics, ultrafast phenomena, ultrasensitive diagnostics, lasers, combustion and plasma physics, optical diagnostics, biomedical applications, and in general, atomic and molecular and optical (AMO) Physics. His Mag. rer. nat. degree shows work on optical bistability at the University of Innsbruck, Austria, with guidance by Dr Peter Zoller. The PhD degree studies are on the subject of polarization spectroscopy and magnetically induced switching at the University of Otago, Dunedin, New Zealand, with guidance by Drs Wes Sandle and Rob Ballagh. He also holds the Dr rer. nat. degree in physics from the University of Innsbruck, Austria. Since 1987, his work has encompassed experimental, theoretical and computational research together with teaching, service, and outreach at the Center for Laser Applications at The University of Tennessee Space Institute, Tullahoma, Tennessee, USA.

James O Hornkohl

Photo of James O and Jeri Hornkohl courtesy of UTSI photo albums: L Horton

The contributions of James Hornkohl or 'Jim' encompass the spectroscopy of diatomic molecules, the application of such spectroscopy in diagnosis of combustion, plasmas, rocket propulsion and related problems. Especially significant are his support of student theses and dissertations, including the application of numerical methods in analysis of experiments. Moreover, his help was greatly appreciated by his collaborators in the design of computational and experimental methods to record digital data. During his last 30 years prior to his death on Februray 7, 2017, Jim had been strongly engaged in the description of the very details on diatomic spectroscopy. The challenge to Jim's work had been the prediction of spectra with a focus on diatomic spectroscopy. The aim of his lifetime work was the design of an algorithm for the prediction and fitting of computed to measured molecular spectra and to provide inferences of parameters such as excitation temperature. The means to accomplish goals for various transitions of diatomic molecules are the consistent application of the standard quantum theory of angular momentum.

Part I

Fundamentals of the diatomic molecule

Christian G Parigger and James O Hornkohl

Chapter 1

Primer on diatomic spectroscopy

1.1 Overview

This book describes how one uses quantum mechanics to predict spectra of diatomic molecules in their gaseous state. The two most important attributes of a spectral line are its position in the electromagnetic spectrum and the strength with which the molecule can interact with the radiation field to produce spectral lines. Thus, a book that discusses the calculation of positions and intensities of spectral lines of a diatomic molecule equally communicates the application of quantum theory to the diatomic molecule.

The theoretically convenient measure of spectral line position is its vacuum wavenumber $\tilde{\nu}_{u\ell}$ which is the difference between the upper term T_u (i.e. upper energy eigenvalue expressed in the units of cm^{-1}) and the lower term T_ℓ,

$$\tilde{\nu}_{u\ell} = T_u - T_\ell. \tag{1.1}$$

In the optical region, the term difference corresponds to a specific color. However, experiments usually measure the wavelength positions in a laboratory setting at standard ambient temperature and pressure. For typical laser spectroscopy investigations of, say, optical emission spectroscopy subsequent to generation of a laser spark, spectral resolutions of the instrument spectrometer + detector amount to 0.1–0.01 nm, rarely to 0.001 nm or 1 pm. At the wavelength, λ, of 400 nm, a spectral resolution, $\Delta\lambda$, of better than 1 pm corresponds to a resolving power, R,

$$R = \lambda/\Delta\lambda \geqslant 400\,000, \tag{1.2}$$

or a wavenumber resolution of better than 0.05 cm^{-1}. The spectral resolution of diatomic molecular data computed in this book is better than 0.05 cm^{-1}. For laser-induced optical breakdown experiments, one recent application of diatomic molecular spectroscopy, resolving powers are of the order of 4000–10 000. For high-resolution, absorption measurements of stellar astrophysical objects, resolving powers of the order of 40 000 are quite common.

The theoretically most convenient measure of a molecule's ability to interact with electromagnetic radiation is its Condon and Shortley [1] line strength, $S_{u\ell}$, that describes transitions between an upper, u, and a lower level, ℓ. The line strength represents a summation over individual states that comprise upper and lower levels. Both the vacuum wavenumber $\tilde{\nu}_{u\ell} = \tilde{\nu}_{\ell u}$ and the line strength $S_{u\ell} = S_{\ell u}$ are symmetric with regard to the upper and lower levels. Also, the symbols u and ℓ each represent a collection of quantum numbers. In diatomic spectroscopy upper state quantum numbers are normally denoted with a single prime while lower states are denoted with the absence of a prime or a double prime. The absence of a double prime has become the standard way of denoting a lower state diatomic quantum number.

1.2 Reversed angular momentum

Historically, the reversed-angular-momentum (RAM) methodology has been successful for prediction of diatomic spectra without the use of modern digital computers. The RAM method establishes a reduced set of basis states, in other words, works with an *a priori* approximation. Sets of rules are introduced when applying a transformation to a molecular-fixed from the laboratory-fixed coordinate system. These rules utilize a supposed reversal of sign in the application of quantum mechanical angular momentum algebra. This section provides a brief historic account of the challenges associated with the RAM method.

The reversed angular momentum approach is mentioned first in an article on the quantization question of the asymmetric top [2]. Klein writes in the introduction that the paper might be of interest for methods of quantization. The reversed sign is introduced for the equations of the components of angular momentum in the molecular-fixed coordinate system in order to obtain agreement with the well-established classical equations for the symmetric top. Conversely, the application of the standard, laboratory-fixed angular momentum equations would lead to the wrong classical result. The article also makes reference to canonical conjugate Euler angles that are interpreted as references to dual space.

The RAM methodology is embraced by Van Vleck in his work on the coupling of angular momentum vectors in molecules [3]. Notably, Sir Harold Kroto communicates in his acceptance lecture for the 1996 Nobel Prize in Chemistry, 'Symmetry, Space, Stars and C_{60}' [4], the prologue *'Symmetry, the Key to the Theory of Everything'*. With reference to the RAM work, Sir Kroto quotes Van Vleck: *Practically every-one (!) knows that the components of total angular momentum (NB the angular momentum operator is usually denoted by J and the associated quantum number by j) of the molecule relative to the axes [x, y, z] fixed in space satisfy the commutation relation of the form*

$$J_xJ_y - J_yJ_x = iJ_z \tag{1.3}$$

Klein discovered the rather surprising fact that when total angular momentum is referred to axes mounted in the molecule which we will denote by $[x', y', z']$ *the sign of i in the commutation relation is reversed i.e.*

$$J_{x'}J_{y'} - J_{y'}J_{x'} = -iJ_{z'} \tag{1.4}$$

Sir Kroto goes on to say: *Does practically everyone know this?—I wondered whether to check this claim out by asking everyone on the main street in Brighton whether they did. I hardly knew—or more accurately—really understood the first relation, let alone the second. However I did know that angular momentum was quantised and governed by the fundamental relations*

$$\langle j|J^2|\rangle = \hbar j(j+1) \tag{1.5}$$

$$M_J = -j\ldots+j \tag{1.6}$$

which means that J has $2j+1$ *possible orientations, and*

$$\Delta j = 0, \pm 1 \tag{1.7}$$

which indicates that when a transition occurs, j may only change by one unit or on occasion remain unchanged.' Previously, in 1975 and then in 1992, Sir Kroto discusses the molecule-fixed angular momentum following Van Vleck [3] leading to the reversed angular momentum equations in his Nobel laureate lecture [4] and in his book on molecular rotation spectra [5].

However, accurate review shows that there is no reversal of the sign when moving from a lab-fix to a molecule-fixed coordinate system, in other words, there is no mathematical support of the reversed sign. Sustenance of the angular momentum equations can be explained as follows: in terms of classical mechanics, reversal of motion occurs as one goes from a rotating system to a fixed system, or vice versa. For example, motion reversal can be experienced by looking at the surroundings while on a rotating merry-go-round versus observing the rotation in the fixed reference frame. The quantum mechanical implementation of motion- or time-reversal is changing the sign and taking the conjugate complex, leading to preservation of the sign. Reference to dual space would confuse things as clearly the standard angular momentum operator equations are not affected by a transformation from lab-fixed to molecule-fixed coordinates, see appendix A.

However, a reasonably concise treatment shows preservation of the commutator relations under a unitary transformation: Consider the operators A, B, and C which satisfy the commutation formula

$$AB - BA = iC \tag{1.8}$$

and subject these three operators to the unitary transformation U. That is,

$$A' = U^{\dagger} A U \tag{1.9a}$$

$$A = U A' U^{\dagger} \tag{1.9b}$$

with similar equations holding for B and C. Then

$$AB - BA = U\,A'U^{\dagger}U\,B'U^{\dagger} - U\,B'U^{\dagger}\,U\,A'U^{\dagger} \tag{1.10a}$$

$$= U\,A'B'U^{\dagger} \tag{1.10b}$$

$$= iC \tag{1.10c}$$

$$iU^{\dagger}\,CU = A'B' \tag{1.10d}$$

$$iC' = A'B' - B'A'. \tag{1.10e}$$

The above result, e.g. see Davydov [6], holds for all commutators including those for angular momentum. Thus,

$$J_{x'}J_{y'} - J_{y'}J_{x'} = i\,J_{z'} \tag{1.11a}$$

$$J_{y'}J_{z'} - J_{z'}J_{y'} = i\,J_{x'} \tag{1.11b}$$

$$J_{z'}J_{x'} - J_{x'}J_{z'} = i\,J_{y'} \tag{1.11c}$$

In summary, the RAM method is not at all utilized in this book for the computation of diatomic molecular spectra. RAM is avoided due in part to not needing approximations with the availability of modern digital computers and due in part to the mathematical inconsistency of the supposed change of sign as implied by the 'reversed angular momentum' descriptive nomenclature.

1.3 Exact diatomic eigenfunction

An exact expression of the diatomic eigenfunction is essential for prediction of spectra. The major difference between this book and other treatments of the diatomic molecule is the use of the Wigner–Witmer diatomic eigenfunction [7] in place of invoking the Born–Oppenheimer approximation [8] from the very beginning of a theory description. In the Wigner–Witmer approach, angular coordinates are exactly separated from the electronic vibrational coordinates. In this book, the Wigner–Witmer eigenfunction is employed for computation of the vacuum wavenumbers and the rotational line strengths. If one were to instead adopt the Born–Oppenheimer approximation, then the rotational line strengths are labeled as Hönl–London factors. The Born–Oppenheimer approximation breaks the electronic–vibrational strength into electronic and vibrational parts that correspond to r-centroids and Franck–Condon factors, and both may be functions of the total angular momentum in the upper and lower levels.

The expression *spectroscopic accuracy* refers to the accuracy with which line position measurements can be performed. Whereas wavelength measurements having an accuracy of 1 part per million are routinely performed, achieving an accuracy of 1 part per hundred in the measurement of relative intensities of a group of spectral lines is fully adequate for many purposes. Thus, one may elect to directly use the Born–Oppenheimer approximation for many practical calculations of molecular line intensity, namely, approximating the diatomic eigenfunction as a product of electronic, vibrational, and rotational factors. However, the Born–Oppenheimer approximation cannot produce diatomic term values with spectroscopic accuracy without generalization. To achieve spectroscopic accuracy within the Born–Oppenheimer approximation, one must include sums over the many electronic states of the molecule and sums over the many vibrational states of each electronic state. Van Vleck transformations [9] or other mathematical procedures reduce the dimension of the Hamiltonian matrix prior to numerically diagonalization [10–15].

In this book, only one diatomic selection rule is used. A spectral line, i.e. a term difference, is allowed if the angular momentum part of its line strength is non-vanishing. However, a modification of the line strength computation is required if the diatomic molecule in question is homonuclear, i.e. the two nuclei are identical. Unresolved hyperfine structure in the spectrum of a homonuclear molecule causes states of positive parity and negative parity to have different nuclear spin statistical weights, g_+ and g_-. If the nuclear spin is zero, either g_+ or g_- will be zero. Thus exchange symmetry, the symmetry associated with the exchange of identical particles, rigorously forbids certain spectral line even when the rotational line strength is non-zero. However, if the rotational line strength factor vanishes, the spectral line is rigorously forbidden.

1.4 Computation of diatomic spectra

The required steps for computation of spectra can be summarized as follows:

- An angular momentum coupling model must be chosen because angular momentum theory does not tell us how the total angular momentum is formed from the orbital and spin momenta.
- The eigenfunctions for everything in the system except the total angular momentum are computed.
- With eigenfunctions obtained in the previous step and the chosen angular momentum coupling model, upper and lower Hamiltonians are computed and diagonalized.
- From the orthogonal matrices that diagonalize the upper and lower Hamiltonians, the line strengths are computed for various possible types of transitions, e.g. electric dipole, magnetic dipole, electric quadrupole, etc. Typically, one knows precisely what type of transition dominates in the spectrum, but this is not invariably the case.
- The non-vanishing of the rotational angular momentum part of the line strength selects the subset of allowed spectral lines from the computed term differences.

Consequently, the minimal information required for computation of a spectrum includes selected term differences $\tilde{\nu}_{u\ell}$ and the computed line strengths $S_{u\ell}$. Description of a diatomic molecule having N electrons and residing in field free space requires $3N + 6$ spatial or angular coordinates, the time t, N electronic spin variables and two nuclear spin variables. In the case of the diatomic molecule, the only exactly separable variables are the time t, the coordinates of the total mass, and three Euler angles which describe the total angular momentum. The Wigner–Witmer diatomic eigenfunction provides the exact separation of three Euler angles, but $3N$ internal spatial coordinates and the numerous spins remain. Unless the number of electrons N is very small, the diatomic problem remains unsolvable with spectroscopic accuracy because there are $3N$ independent variables that cannot be treated with mathematical exactness.

Despite the challenges mentioned in the previous paragraph, one can, with two stringent caveats, apply the above algorithm to the diatomic molecule. The first caveat is that one must have extensive experimentally recorded wave number tables, $\tilde{\nu}_{u\ell}^{exp}(J', J)$, versus upper and lower total angular momenta, J' and J, respectively, for many vibrational bands in the spectrum of a molecule of interest. The second caveat is associated with using trial values of semi-empirical molecular parameters for each vibrational level, v, such as B_v, D_v, A_v, λ_v, γ_v ..., etc. One computes term differences, $\tilde{\nu}_{u\ell}(J', J)$, from numerically diagonalized upper and lower Hamiltonians, calculates corrections to the trial values of the parameters from differences $\tilde{\nu}_{u\ell}(J', J) - \tilde{\nu}_{u\ell}^{exp}(J', J)$, and iterate the computations until the errors in the computed line positions are comparable to the estimated errors in the experimental line positions. When successful, this procedure yields working models for the upper and lower Hamiltonians and sets of molecular parameters which predict the measured line positions.

The practical significance of molecular parameters was their appearance in term value equations, semi-empirical equations with which one can compute the upper T_u and lower T_ℓ terms and thereby the vacuum wavenumber $\tilde{\nu}_{u\ell}$. Herzberg [16] gives many examples of term value equations, but note that when Herzberg wrote his book, numerical diagonalization of thousands of matrices was impractical. The current significance of the molecular parameters is that they can be used to compute diatomic Hamiltonian matrix representations in one of Hund's bases.

In this book, computation of $\tilde{\nu}_{u\ell}(J', J)$ and $S_{u\ell}(J', J)$ is based upon the Wigner–Witmer diatomic eigenfunction instead of the eigenfunction associated with the Born–Oppenheimer approximation, but computations of the electronic–vibrational strengths utilize separation of electronic from vibrational contributions familiar from the Born–Oppenheimer approximation.

References

[1] Condon E U and Shortley G 1953 *The Theory of Atomic Spectra* (Cambridge: Cambridge University Press)
[2] Klein O 1929 Zur Frage der Quantelung des asymmetrischen Kreisels *Z. Phys.* **58** 730
[3] Van Vleck J H 1951 The coupling of angular momentum vectors in molecules *Rev. Mod. Phys.* **23** 213
[4] Kroto H W 1996 Symmetry, space, stars and C_{60} (accessed January 19, 2016) (http://www.nobelprize.org/prizes/chemistry/1996/kroto/facts/)
[5] Kroto H W 1992 *Molecular Rotation Spectra* (New York: Dover)
[6] Davydov A S 1965 *Quantum Mechanics* (Oxford: Pergamon)

[7] Wigner E and Witmer E E 1928 *Z. Phys.* **51** 859
Hettema H 2000 *Quantum Chemistry: Classic Scientific Papers* (Singapore: World Scientific) p 287
[8] Born M and Oppenheimer R 1927 *Ann. Phys.* **84** 457
Hettema H 2000 *Quantum Chemistry: Classic Scientific Papers* (Singapore: World Scientific) p 1
[9] Kemble E C 2005 *The Fundamental Principles of Quantum Mechanics* (Mineola, NY: Dover)
[10] Zare R N, Schmeltekopf A L, Harrop W J and Albritton D L 1973 *J. Mol. Spectrosc.* **46** 37
[11] Bunker P R and Jensen P 1998 *Molecular Symmetry and Spectroscopy* 2nd edn (Ottawa: NRC)
[12] Brown J M and Carrington A 2003 *Rotational Spectroscopy of Diatomic Molecules* (Cambridge: Cambridge University Press)
[13] Lefebvre-Brion H and Field R W 2004 *The Spectra and Dynamics of Diatomic Molecules* (New York: Elsevier/Academic)
[14] Watson J K G 2005 Different forms of effective Hamiltonians *J. Mol. Spectrosc.* **103** 3283
[15] Field R W, Baraban J H, Lipoff S H and Beck A R 2011 *Handbook of High-resolution Spectroscopy* ed M Quack and F Merkt (Hoboken, NJ: Wiley)
[16] Herzberg G 1950 *Molecular Spectra and Molecular Structure I. Spectra of Diatomic Molecules* 2nd edn (New York: Van Nostrand Reinhold)

IOP Publishing

Christian G Parigger and James O Hornkohl

Chapter 2

Line strength computations

2.1 Introduction

The quantum theory description of optical transitions in diatomic molecules uses the Condon and Shortley [1] line strength, $S_{u\ell}$,

$$S_{u\ell} = \sum_{m_u} \sum_{m_\ell} \sum_{k=-q}^{q} |\langle \ell | T_k^{(q)} | u \rangle|^2, \tag{2.1}$$

where $T_k^{(q)}$ is the kth component of the tensor operator $\mathbf{T}^{(q)}$ of degree q, e.g. electric dipole, magnetic dipole, electric quadrupole, etc. The tensor operator describes the type of interaction between the molecule and the radiation field. The basic idea behind summations in the line strength formula is to include all transitions that produce the same spectral line. In field free space, the summations are over the upper m_u and lower m_ℓ magnetic quantum numbers, and the k components of the tensor operator.

The line position is described by the vacuum wavenumber $\tilde{\nu}_{u\ell}$ which is the difference between the upper term T_u, i.e. upper energy eigenvalue expressed in the units of cm^{-1}, and the lower term T_ℓ,

$$\tilde{\nu}_{u\ell} = T_u - T_\ell. \tag{2.2}$$

The vacuum wavenumber $\tilde{\nu}_{u\ell} = \tilde{\nu}_{\ell u}$ and the line strength $S_{u\ell} = S_{\ell u}$ are symmetric with regard to the upper and lower levels. Also, the symbols u and ℓ each represent a collection of quantum numbers. For a diatomic molecule, the symbol u represents the collection n', v', J', and M' while ℓ represents n, v, J, and M in which v is the vibrational quantum number, J is the total angular momentum quantum number, M is the quantum number for the z-component of $\mathbf{J}$, and n represents the collection of all other required quantum numbers. In diatomic spectroscopy, upper state quantum numbers are normally denoted with a single prime while lower states are denoted with the absence of a prime or a double prime. The absence of a double prime has become the standard way of denoting a lower state diatomic quantum number.

The line position is described by the vacuum wavenumber $\tilde{\nu}_{u\ell}$ which is the difference between the upper term T_u, (i.e. upper energy eigenvalue expressed in the units of cm^{-1}) and the lower term T_ℓ,

$$\tilde{\nu}_{u\ell} = T_u - T_\ell. \tag{2.3}$$

In concise summary, this book is about how to compute the vacuum wavenumbers,

$$\tilde{\nu}(n'v'J', nvJ) = T(n'v'J') - T(nvJ) \tag{2.4}$$

and line strengths,

$$S(n'v'J', nvJ) = \sum_{M'=-J}^{J'} \sum_{M=-J}^{J} \sum_{k=-q}^{q} \left| \left\langle nvJM \middle| T_k^{(q)} \middle| n'v'J'M' \right\rangle \right|^2, \tag{2.5}$$

for diatomic molecules. The upper and lower states are assumed to have a degeneracy (statistical weight) of $2J' + 1$ and $2J + 1$. This implies that the molecule is free from any static fields. This also explains why M' and M do not appear in the equation for the vacuum wavenumber $\tilde{\nu}$, and why there are sums over M' and M in the equation for the line strength $S(n'v'J', nvJ)$. Noteworthy, the Wigner–Eckart theorem separates the transition matrix element into two parts one of which depends entirely on angular momentum,

$$\left\langle nvJM \middle| T_k^{(q)} \middle| n'v'J'M' \right\rangle = \frac{C_{J'M'qk}^{JM}}{\sqrt{2J+1}} \langle nvJ || \mathbf{T}^{(q)} || n'v'J' \rangle. \tag{2.6}$$

The exact Wigner–Witmer [2] diatomic eigenfunction,

$$\begin{aligned} &\langle \rho, \zeta, \chi, \mathbf{r}_2, \dots, \mathbf{r}_N, r, \theta, \phi \,| \mathcal{R}(\alpha, \beta, \gamma) | nJM \rangle \\ &= \sum_{\Omega=-J}^{J} \langle \rho, \zeta, \mathbf{r}'_2, \dots, \mathbf{r}'_N, r \,| nv \rangle \, D_{M\Omega}^{J^*}(\phi, \theta, \chi), \end{aligned} \tag{2.7}$$

further elaborated and derived in chapter 4, provides the primary theory model used in this book as opposed to the Born–Oppenheimer [3] approximation, and this is the major difference between contents in this book and other treatments of the diatomic molecule. In the Wigner–Witmer eigenfunction, the angular coordinates ϕ, θ, and χ are exactly separated from the electronic vibrational coordinates $\rho, \zeta, \mathbf{r}'_2, \dots, \mathbf{r}'_N, r$. Coordinates r, θ, and ϕ are the spherical coordinates of the internuclear vector $\mathbf{r}(r, \theta, \phi)$ with the coordinate origin at the center of mass of the two nuclei. Coordinates ρ, ζ, and χ are the cylindrical coordinates $\mathbf{r}'_1(\rho, \zeta, \chi)$ of one of the N electrons of the mutual nuclei and electron configuration. Rotation of the mutual configuration is accomplished by arbitrarily choosing electron 1 in rotated, primed coordinates obtained from unprimed coordinates through coordinate rotations ϕ, θ, and χ. A primed quantum number denotes the upper state. A primed coordinate denotes a rotated coordinate.

In this book, the Wigner–Witmer eigenfunction is used for computation of the vacuum wavenumber and the unitless rotational line strength factor, $S(J', J)$, the angular momentum part of the diatomic line strength, viz. Hönl–London factor in descriptions that utilize the Born–Oppenheimer approximation from the very start of computing diatomic spectra,

$$S(n'v'J', nvJ) = S_{\text{ev}}(n'v'J', nvJ)\, S(J', J), \tag{2.8}$$

in which $S_{\text{ev}}(n'v'J', nvJ)$ is the electronic–vibrational strength of the diatomic molecule and which carries the units of line strength (C^2 m^2 for electric dipole transitions). The electronic–vibrational strength $S_{\text{ev}}(n'v'J', nvJ)$ is mostly controlled by the electronic and vibrational eigenfunctions and is a weaker function of the angular momentum quantum numbers J' and J. Typically, if the diatomic molecule is not a hydride, the variation of electronic–vibrational strength $S_{\text{ev}}(n'v'J', nvJ)$ can often be ignored and it is denoted $S_{\text{ev}}(n'v', nv)$. Subsequently, the reduced Born–Oppenheimer approximation breaks the electronic–vibrational strength into electronic and vibrational parts

$$S_{\text{ev}}(n'v', nv) \approx |a_0 + a_1\bar{r}_1(v', v) + a_2\bar{r}_2(v', v) + \ldots|^2\; q(v', v) \tag{2.9}$$

in which

$$\bar{r}_n(v', v) = \int_0^\infty \psi_{v'}(r)\, r^n\, \psi_v(r)\, dr \Big/ \int_0^\infty \psi_{v'}(r)\, \psi_v(r)\, dr, \tag{2.10}$$

$$q(v', v) = \left| \int_0^\infty \psi_{v'}(r)\, \psi_v(r)\, dr \right|^2 \tag{2.11}$$

The upper and lower vibrational eigenfunctions, $\psi_{v'}(r)$ and $\psi_v(r)$, are not elements of the same vector space (e.g. $\langle v'|v\rangle \neq \delta_{v',v}$) because they are eigenfunctions for different potential energy wells. For those cases for which the variation of $S_{\text{ev}}(n'v'J', nvJ)$ with J' and J cannot be ignored, the vibrational eigenfunctions become functions of angular momentum, and the Franck–Condon factors $q(v', v)$ and r-centroids $\bar{r}_n(v', v)$ acquire angular momentum dependence to become $q(v'J', vJ)$ and $\bar{r}_n(v'J', vJ)$.

The expression *spectroscopic accuracy* refers to the accuracy with which line position measurements can be performed. Whereas wavelength measurements having an accuracy of 1 part per million are routinely performed, achieving an accuracy of 1 part per hundred in the measurement of relative intensities of a group of spectral lines is fully adequate for many purposes. For measurement accuracies of relative intensities of the order of 1 part per hundred, the standard Born–Oppenheimer approximation,

$$\Psi_{nvJM} \approx \psi_n\, \psi_v\, \psi_{JM}, \tag{2.12}$$

is adequate for many practical calculations of molecular spectra. However, the Born–Oppenheimer approximation (2.12) cannot produce diatomic term values with spectroscopic accuracy. To achieve spectroscopic accuracy using the Born–Oppenheimer approximation, one would have to replace equation (2.12) with

$$\Psi_{nvJM} \approx \sum_{n} \sum_{v_n} \psi_n \, \psi_{v_n} \, \psi_{JM}, \tag{2.13}$$

in which the n sum is over the many electronic states of the molecule and the v_n sums are over the many vibrational states of each electronic state.

2.2 Idealized computation of spectra

What if mathematics were no impediment and we could actually perform the procedures described in quantum mechanics textbooks required for the solution of a quantum system having bound states. For example, what if we were given a complete basis $|a\rangle$ with which we could compute a matrix representation of the Hamiltonian H having matrix elements

$$H_{ij} = \langle a_i | H | a_j \rangle. \tag{2.14}$$

In principle, the Hamiltonian matrix is diagonal with dimensions $(2J + 1) \times (2J + 1)$, where J is the total angular momentum quantum number. Angular momentum theory tells us that the total angular momentum $\mathbf{J}$ is the quantity associated with the empirical observation that field free space is isotropic, but there are two kinds of angular momenta, orbital $\mathbf{N}$ and spin $\mathbf{S}$, and angular momentum theory does not tell us how the orbital and spin momenta are coupled to form the total $\mathbf{J}$. We must form a mathematically complete basis for an assumed coupling scheme, e.g. ℓs or jj for atoms or one of Hund's models for molecules. The physical state $|nJM\rangle$ must be expressed as a sum of our basis states $|a_i\rangle$,

$$|nJM\rangle = \sum_i |a_i\rangle \, \langle a_i | nJM \rangle. \tag{2.15}$$

The result is still mathematically exact because the basis $|a_i\rangle$ is complete, but two practical problems are caused by using basis functions instead of the eigenfunctions for the system. First, the Hamiltonian matrix computed using the basis functions is no longer diagonal. Second, the Hamiltonian matrix has become larger. For example, if $J = 10$ and 10 basis functions are required, the Hamiltonian matrix dimensions are now $(210) \times (210)$, i.e. $(2J + 1) \times 10 = 210$. The quantum mechanics of diatomic molecules played an important part in the early development of quantum theory, but at the time numerical or analytical diagonalization of a 4×4 matrix was a serious challenge, and this remained true until roughly 1960. Today, numerical diagonalization of much larger matrices can be accomplished routinely.

Numerical diagonalization of the upper H_u and lower H_ℓ Hamiltonians gives the upper and lower energy eigenvalues, i.e. upper and lower terms T_u and T_ℓ,

$$T_\mathrm{u} = \tilde{U}_\mathrm{u} \, H_\mathrm{u} \, U_\mathrm{u} \tag{2.16a}$$

$$T_\ell = \tilde{U}_\ell \, H_\ell \, U_\ell \tag{2.16b}$$

and from equation (2.2) we find the line position $\tilde{\nu}_{\mathrm{u}\ell}$. The tilde on a matrix (unrelated to the tilde on the vacuum wavenumber $\tilde{\nu}$) denotes the transpose of that matrix. The

Hamiltonians are assumed to be real symmetric matrices. Thus, the matrices $U_{\rm u}$ and U_ℓ are real unitary matrices. That is, $U_{\rm u}$ and U_ℓ are orthogonal matrices,

$$U_{\rm u}^{-1} = \tilde{U}_{\rm u}, \tag{2.17a}$$

$$U_\ell^{-1} = \tilde{U}_\ell. \tag{2.17b}$$

The elements of U_ℓ are the coefficients from equation (2.15),

$$U_{ij}^{(\rm u)} = \langle a_i|(n'J'M')_j\rangle, \tag{2.18a}$$

$$U_{nm}^{(\ell)} = \langle a_n|(nJM)_m\rangle. \tag{2.18b}$$

In software, a two-dimensional array has two indices, say i and j. In the above equation, the replacement of quantum numbers nJM with the single software index j is an example of something that often happens in software pertaining to applied quantum mechanics.

The line strength $S_{{\rm u}\ell}$ can be computed from the matrices $U_{\rm u}$ and U_ℓ that diagonalized the upper and lower Hamiltonians,

$$S_{nj}^{({\rm u}\ell)} = \sum_{k=-q}^{q} \sum_{m=1}^{\mathcal{N}_\ell} \sum_{i=1}^{\mathcal{N}_{\rm u}} \tilde{U}_{nm}^{(\ell)} \left\langle m \middle| T_k^{(q)} \middle| i \right\rangle U_{ij}^{(\rm u)} \tag{2.19a}$$

where $\mathcal{N}_{\rm u}$ is the dimension of the square upper Hamiltonian matrix $H_{\rm u}$ and $\mathcal{N}_\ell$ is the dimension of H_ℓ.

In equations (2.16), the matrices $T_{\rm u}$ and T_ℓ of the upper and lower terms are diagonal square matrices of dimensions $\mathcal{N}_{\rm u}$ and $\mathcal{N}_\ell$ (in software, one-dimensional arrays). The total number of term differences, i.e. potential spectral lines, is $\mathcal{N}_{\rm u} \times \mathcal{N}_\ell$, but this number of term differences is much larger than the number of actual spectral lines. Selection rules determine which of the term differences represents a real spectral line. In this book, only one diatomic selection rule is used. A spectral line, i.e. a term difference, is allowed if the angular momentum part of its line strength, $S(J', J)$ in equation (2.8), is non-vanishing.

References

[1] Condon E U and Shortley G 1953 *The Theory of Atomic Spectra* (Cambridge: Cambridge University Press)

[2] Wigner E and Witmer E E 1928 *Z. Phys.* **51** 859
Hettema H 2000 *Quantum Chemistry: Classic Scientific Papers* (Singapore: World Scientific) p 287

[3] Born M and Oppenheimer R 1927 *Ann. Phys.* **84** 457
Hettema H 2000 *Quantum Chemistry: Classic Scientific Papers* (Singapore: World Scientific) p 1

Chapter 3

Framework of the Wigner–Witmer eigenfunction (WWE)

In this book, the eigenfunction $\Psi_{nJM}(\mathbf{r}_1, \mathbf{r}_2, \ldots, \mathbf{r}_N)$ is viewed as the projection of the abstract state $|nJM\rangle$ into coordinate space,

$$\Psi_{nJM}(\mathbf{r}_1, \mathbf{r}_2, \ldots, \mathbf{r}_N) \equiv \langle \mathbf{r}_1, \mathbf{r}_2, \ldots, \mathbf{r}_N\ |nJM\rangle. \tag{3.1}$$

Quantum theory postulates the existence of the state $|nJM\rangle$. Schrödinger's eigenfunction reveals how this state manifests itself in the world of calculus, differential equations, and matrix algebra. The quantum numbers here are J for the total angular momentum $\mathbf{J}$, M for the z component of $\mathbf{J}$, and the symbol n representing all other required quantum numbers.

It is the nature of angular momentum that the total $\mathbf{J}$ [actually, the total squared whose eigenvalue is $J(J + 1)$] and *only one component*, conventionally taken to be the z component, are constants of the motion. That is, $\mathbf{J}^2$ and J_z commute with the Hamiltonian operator and other operators composing the complete set of commuting observables (CSCO), but no other components of $\mathbf{J}$ are constants of the motion.

Angular momentum theory gives us the standard angular-momentum representation [1], $|JM\rangle$, in which

$$\mathbf{J}^2|JM\rangle = J(J+1)\ |JM\rangle \tag{3.2a}$$

$$J_z|JM\rangle = M\ |JM\rangle \tag{3.2b}$$

$$J_\pm = J_x \pm i\, J_y \tag{3.2c}$$

$$J_\pm|JM\rangle = \sqrt{J(J+1) - M(M \pm 1)}\ |J, M \pm 1\rangle \tag{3.2d}$$

$$= \sqrt{(J \mp M)(J \pm M + 1)}\ |J, M \pm 1\rangle \tag{3.2e}$$

doi:10.1088/978-0-7503-1989-8ch3

$$= C_{\pm}(JM)\,|J,\, M \pm 1\rangle \tag{3.2f}$$

The homogeneity of time and space and the isotropy of space permit one to perform coordinate translations and rotations in an attempt to simplify the mathematics of the problem by separating some of the variables, that is, by breaking the eigenfunction into a product of independent functions. Each coordinate rotation changes the direction of the z-axis giving a new operator $J_{z'}$ requiring one to work with a new $|J\Omega\rangle$ standard representation of angular momentum,

$$J_{z'}|J\Omega\rangle = \Omega|J\Omega\rangle \tag{3.3a}$$

$$J'_{\pm}|J\Omega\rangle = C_{\pm}(J\Omega)\,|J,\, \Omega \pm 1\rangle J \tag{3.3b}$$

in which Ω is the new magnetic quantum number for the new z' axis. Of particular importance is the ability to move between two standard representations,

$$|J\Omega\rangle = \sum_{M=-J}^{J} |JM\rangle\,\langle JM\,|J\Omega\rangle, \tag{3.4}$$

$$|JM\rangle = \sum_{\Omega=-J}^{J} |J\Omega\rangle\,\langle J\Omega\,|JM\rangle. \tag{3.5}$$

Application of the rotation operator [2], $\mathcal{R}(\alpha, \beta, \gamma)$, on the eigenfunction Ψ_{nJM} gives a new mathematical function $\Phi_{nJ\Omega}$,

$$\mathcal{R}(\alpha, \beta, \gamma)\,\Psi_{nJM}(\mathbf{r}_1, \mathbf{r}_2, \ldots, \mathbf{r}_{\mathcal{N}}) = \Phi_{nJ\Omega}(\mathbf{r}'_1, \mathbf{r}'_2, \ldots, \mathbf{r}'_{\mathcal{N}}) \tag{3.6}$$

where primes denote rotated coordinates. This equation describes coordinate rotation, not mechanical rotation. The Euler angles α, β, and γ are parameters of coordinate rotation, not angles describing mechanical rotation of mass and charge. The eigenfunctions $\Psi_{nJM}(\mathbf{r}_1, \mathbf{r}_2, \ldots, \mathbf{r}_{\mathcal{N}})$ and $\Phi_{nJ\Omega}(\mathbf{r}'_1, \mathbf{r}'_2, \ldots, \mathbf{r}'_{\mathcal{N}})$ describe the exact same physical state and differ only because they are expressed in different coordinate systems.

Using Dirac notation, we apply $\mathcal{R}\mathcal{R}^{\dagger} = 1$ (because $\mathcal{R}$ is unitary) to the eigenfunction $\Psi_{nJM}(\mathbf{r}_1, \mathbf{r}_2, \ldots, \mathbf{r}_{\mathcal{N}})$,

$$\begin{aligned}
\langle \mathbf{r}_1, \mathbf{r}_2, \ldots, \mathbf{r}_{\mathcal{N}}\,|nJM\rangle &= \langle \mathbf{r}_1, \mathbf{r}_2, \ldots, \mathbf{r}_{\mathcal{N}}\,|\mathcal{R}(\alpha, \beta, \gamma)\,\mathcal{R}^{\dagger}(\alpha, \beta, \gamma)|JM\rangle \\
&= \sum_{\Omega=-J}^{J} \langle \mathbf{r}_1, \mathbf{r}_2, \ldots, \mathbf{r}_{\mathcal{N}}\,|\mathcal{R}(\alpha, \beta, \gamma)|nJ\Omega\rangle \\
&\qquad \times \langle J\Omega\,|\mathcal{R}^{\dagger}(\alpha, \beta, \gamma)|JM\rangle \\
&= \sum_{\Omega=-J}^{J} \mathbf{r}'_1, \mathbf{r}'_2, \ldots, \mathbf{r}'_{\mathcal{N}}\,|nJ\Omega\rangle\,\langle JM\;\mathcal{R}(\alpha, \beta, \gamma)\,|J\Omega\rangle^{*},
\end{aligned} \tag{3.7}$$

to yield

$$\langle \mathbf{r}_1, \mathbf{r}_2, \ldots, \mathbf{r}_N | nJM \rangle = \sum_{\Omega=-J}^{J} \mathbf{r}'_1, \mathbf{r}'_2, \ldots, \mathbf{r}'_N | nJ\Omega \rangle \, D^{J^*}_{M\Omega}(\alpha, \beta, \gamma). \tag{3.8}$$

The above result shows that the rotation matrix element or Wigner D-function is the mathematical tool one uses to perform transformations between two standard angular momentum representations which differ only because of coordinate rotation.

Equation (3.8) strictly describes coordinate rotation, not mechanical rotation. However, the simple geometry of the diatomic molecule gives one a way to use (3.8) to describe the physical rotation of the diatomic molecule. This is possible because the mechanical rotations ϕ, θ, and χ of the molecule duplicate the coordinate rotations α, β, and γ.

The eigenfunction describing the internal motions in a diatomic molecule having N electrons will require the electronic coordinates $\mathbf{r}_1, \mathbf{r}_2, \ldots, \mathbf{r}_N$ and the internuclear vector $\mathbf{r}$ with the origin at the center of mass of the two nuclei. The internuclear vector $\mathbf{r}$ is expressed in terms of its polar angle θ, azimuthal angle ϕ, and internuclear distance r,

$$x = r \sin\theta \cos\phi \tag{3.9a}$$

$$y = r \sin\theta \sin\phi \tag{3.9b}$$

$$z = r \cos\theta. \tag{3.9c}$$

The coordinate vector of one of the electrons, arbitrary chosen to be electron 1, is expressed in cylindrical coordinates

$$x_1 = \rho \cos\chi, \tag{3.10a}$$

$$y_1 = \rho \sin\chi, \tag{3.10b}$$

$$z_1 = \zeta, \tag{3.10c}$$

where ρ is the distance of the chosen electron from the internuclear vector, ζ is the distance of that electron above or below the plane perpendicular to the internuclear vector and passing through the center of mass of the two nuclei, and χ is mechanical rotation of the chosen electron about the internuclear vector. With these variables, one can write the internal diatomic eigenfunction as

$$\langle \mathbf{r}_1, \mathbf{r}_2, \ldots, \mathbf{r}_N, \mathbf{r} | nvJM \rangle = \langle \rho, \zeta, \chi, \mathbf{r}_2, \ldots, \mathbf{r}_N, r, \theta, \phi | nvJM \rangle. \tag{3.11}$$

The vibrational quantum number v has been extracted from the collection n.

In accord with equation (3.8), coordinate rotation of the right side of equation (3.11) gives

$$\begin{aligned} &\langle \rho, \zeta, \chi, \mathbf{r}_2, \ldots, \mathbf{r}_N, r, \theta, \phi | nvJM \rangle \\ &= \sum_{\Omega=-J}^{J} \langle \rho, \zeta, \chi', \mathbf{r}'_2, \ldots, \mathbf{r}'_N, r, \theta', \phi' | nvJ\Omega \rangle D^{J^*}_{M\Omega}(\alpha, \beta, \gamma) \end{aligned} \tag{3.12}$$

with

$$\phi' = \phi - \alpha, \tag{3.13a}$$

$$\theta' = \theta - \beta, \tag{3.13b}$$

$$\chi' = \chi - \gamma. \tag{3.13c}$$

The rotationally invariant scalars ρ, ζ, and r remain unprimed. Equations (3.13) hold because mechanical rotation ϕ and coordinate rotation α are about the z axis, mechanical rotation θ and coordinate rotation β are about the first intermediate y axis of the full coordinate rotation, and mechanical rotation χ and coordinate rotation γ are about the z' axis. One is at liberty to choose the angles of coordinate rotation. The choices,

$$\alpha = \phi, \tag{3.14a}$$

$$\beta = \theta, \tag{3.14b}$$

$$\gamma = \chi, \tag{3.14c}$$

remove all angular dependence from

$$\langle \rho, \zeta, \chi' = 0, \mathbf{r}_2', \ldots, \mathbf{r}_N', r, \theta' = 0, \phi' = 0 \,|nJ\Omega\rangle \tag{3.15}$$

and the result is the Wigner–Witmer diatomic eigenfunction [3],

$$\begin{aligned}\langle \rho, \zeta, \chi, \mathbf{r}_2, \ldots, \mathbf{r}_N, r, \theta, \phi \,|nvJM\rangle = \sum_{\Omega=-J}^{J} \langle \rho, \zeta, \mathbf{r}_2', \ldots, \mathbf{r}_N', r \,|nv\rangle \\ \times D_{M\Omega}^{J^*}(\phi, \theta, \chi).\end{aligned} \tag{3.16}$$

The electronic–vibrational eigenfunction $\langle \rho, \zeta, \mathbf{r}_2', \ldots, \mathbf{r}_N', r \,|nv\rangle$ is influenced by angular momentum (e.g. by centrifugal stretching) but it is not the total angular momentum eigenfunction. The variables ϕ, θ, and χ are exactly separated in the Wigner–Witmer diatomic eigenfunction. Because the general equation for coordinate rotation (3.8) holds only if J, M, and Ω refer to the total angular momentum, all orbital and spin momenta are included in $D_{M\Omega}^{J^*}(\phi, \theta, \chi)$.

References

[1] Arfken G B, Weber H J and Harris F E 2012 *Mathematical Methods for Physicists, A comprehensive Guide* 7th edn (New York: Academic)

[2] Wolf A A 1969 *Am. J. Phys.* **37** 531

[3] Wigner E and Witmer E E 1928 *Z. Phys.* **51** 859
Hettema H 2000 *Quantum Chemistry: Classic Scientific Papers* (Singapore: World Scientific) p 287

Christian G Parigger and James O Hornkohl

Chapter 4

Derivation of the Wigner–Witmer eigenfunction

This chapter gives a derivation of the Wigner–Witmer diatomic eigenfunction based on symmetries associated with conservation of energy, linear momentum, and angular momentum. An experimental observation recorded at a given time, location, and orientation should be reproducible are all other times, locations, and orientations. Time and space are homogeneous, and space is isotropic. The Wigner–Witmer diatomic eigenfunction [1], i.e., the diatomic eigenfunction in which time, the coordinates of the center of mass, and the coordinates of the total angular momentum have been exactly separated, follows directly from consideration of translational symmetry of time and space and rotational symmetry of space [2]. Application of the time translation operator $\mathcal{U}$, spatial translation operator $\mathcal{T}$, and rotation operator $\mathcal{R}$ to the total eigenfunction yields a separation of time, spatial coordinates of the total linear momentum, and angular coordinates of the total angular momentum.

The system under consideration in this book is a diatomic molecule in the radiation field. This system is conservative. Total energy, total linear momentum, and total angular momentum are constants of the motion. In this chapter, we assume the molecule is in its ground state and the radiation field is empty. That is, in this chapter, we treat a diatomic molecule as a conservative system.

4.1 Outline of the derivation

The eigenfunction for a diatomic molecule having N electrons and, of course, two nuclei depends on the N spatial coordinates $\mathbf{R}_1, \mathbf{R}_2, \ldots, \mathbf{R}_N$ of the electrons, the spatial coordinates $\mathbf{R}_a$ and $\mathbf{R}_b$ of the nuclei and time,

$$\Psi_{nvJM}(\mathbf{R}_1, \mathbf{R}_2, \ldots, \mathbf{R}_N, \mathbf{R}_a, \mathbf{R}_b, t) \equiv \langle \mathbf{R}_1, \mathbf{R}_2, \ldots, \mathbf{R}_N, \mathbf{R}_a, \mathbf{R}_b, t|nvJM\rangle. \quad (4.1)$$

doi:10.1088/978-0-7503-1989-8ch4

J and M again refer to the total angular momentum including all spins, and n refers to all other quantum numbers except the vibrational quantum number v.

The following derivation consists of replacing the coordinates in equation (4.1) with coordinates ρ, ζ, χ, $\mathbf{r}_2$, ..., $\mathbf{r}_N$, r, ϕ, θ, $\mathbf{R}_{\text{CM}}$, τ. The new coordinates are introduced for the purpose of making symmetry transformations. At first, the symmetry transformations merely change the origins of the coordinates (independent variables). These coordinate transformations give new mathematical pictures of a physically unchanged system. The independent variables displayed in equation (4.1) eigenfunctions are each referenced to an arbitrarily chosen origin 0, but there is no absolute origin for time, cartesian coordinates, or angular coordinates. Time and space are homogeneous, and space in isotropic. If the new origins of the independent variables are referenced to the internal dynamics of the molecule, the new independent variables become dynamical variables describing physical motion independent of fictitious absolute origins for time, space, or orientation. The homogeneity of time and space and the isotropy of space allow one to choose new origins for the independent variables. Certain numerical values, e.g. the total energy, total linear momentum, total angular momentum, and the mathematical forms certain important equations, e.g. the Schrödinger equation and the angular momentum commutation formulas, are invariant under the symmetry transformations. If the origins of the new variables are connected to the electrons and nuclei composing a conservative system, invariance of the total energy, linear momentum, and angular momentum under symmetry transformations is converted to their conservation during physical motion. The approach taken here is to minimize initial approximations in the treatment of diatomic molecules in favor of using the tools of modern quantum physics.

4.2 Time translation symmetry

Our discussion of time translation symmetry begins with the time dependent Schrödinger equation,

$$\frac{\partial \Psi(t)}{\partial t} = -\frac{i}{\hbar} H \, \Psi(t). \tag{4.2}$$

For a conservative system, the Hamiltonian H is not a function of time. For a system having bound states, the Hamiltonian H has eigenvalues E,

$$H \, \Psi(t) = E \, \Psi(t). \tag{4.3}$$

The time translation symmetry operator $\mathcal{U}(t, t_0)$ is defined by the effect on the eigenfunction $\Psi(t)$ caused by changing the origin of the time coordinate t from $t = 0$ to some new origin t_0,

$$\mathcal{U}(t, t_0) \, \Psi(t) = \Psi(t - t_0). \tag{4.4}$$

The Taylor series of $\Psi(t - t_0)$,

$$\Psi(t - t_0) = \Psi(t) - t_0\frac{\partial\Psi(t)}{\partial t} + \frac{t_0^2}{2!}\frac{\partial^2\Psi(t)}{\partial t^2} - \frac{t_0^3}{3!}\frac{\partial^3\Psi(t)}{\partial t^3} + \frac{t_0^4}{4!}\frac{\partial^4\Psi(t)}{\partial t^4} + \cdots \quad (4.5)$$

in combination with the result

$$\frac{\partial^k\Psi}{\partial t^k} = \left(\frac{i\,E}{\hbar}\right)^k\Psi(t) \quad k \text{ even} \quad = -\left(\frac{i\,E}{\hbar}\right)^k\Psi(t) \quad k \text{ odd} \quad (4.6)$$

from the time dependent Schrödinger equation (4.2) and equation (4.3) gives

$$\begin{aligned}\Psi(t - t_0) = \Psi(t) &+ \frac{i\,t_0\,E}{\hbar}\Psi(t) + \frac{1}{2!}\left(\frac{i\,t_0\,E}{\hbar}\right)^2\Psi(t) \\ &+ \frac{1}{3!}\left(\frac{i\,t_0\,E}{\hbar}\right)^3\Psi(t) + \frac{1}{4!}\left(\frac{i\,t_0\,E}{\hbar}\right)^4\Psi(t) + \cdots\end{aligned} \quad (4.7)$$

yielding the time translation operator,

$$\mathcal{U}(t, t_0) = \exp\left(\frac{i\,t_0\,E}{\hbar}\right). \quad (4.8)$$

The time translation operator is unitary,

$$\mathcal{U}^{-1}(t, t_0) = \mathcal{U}^{\dagger}(t, t_0) \quad (4.9a)$$

$$= \exp\left(-\frac{i\,t_0\,E}{\hbar}\right). \quad (4.9b)$$

Equation (4.4) can be rewritten as

$$\Psi(t) = \mathcal{U}^{\dagger}(t, t_0)\Psi(t - t_0) \quad (4.10a)$$

$$= \exp\left(-\frac{i\,t_0\,E}{\hbar}\right)\Psi(t - t_0). \quad (4.10b)$$

If the new time origin is not referenced to a physical event in the system, the new origin t_0 becomes just another independent variable. If t_0 is equated to the time at which some event occurred in the system, then $\mathcal{U}^{\dagger}(t, t_0)$ describes the temporal evolution of the system from that event. Equating t_0 to the time at which some event occurred in the system is identical to imposing a boundary condition on the solution of the time-dependent Schrödinger equation. The time-dependent Schrödinger equation is easily solved when the Hamiltonian H is not a function of time and has the eigenvalue E. The time dependent Schrödinger equation becomes the simple differential equation

$$\frac{d\Psi(t)}{\Psi(t)} = -\frac{i}{\hbar}E\,dt \quad (4.11)$$

whose solution is

$$\Psi(t) = \text{const}\,\exp\left(-\frac{i\,E\,t}{\hbar}\right). \tag{4.12}$$

One might use the boundary condition $\Psi(t = 0) = \Psi(0)$ which gives

$$\Psi(t) = \exp\left(-\frac{i\,E\,t}{\hbar}\right)\Psi(0). \tag{4.13}$$

Note that this same result is obtained when t_0 is set equal to t in equation (4.10b).

The alternative boundary condition $\Psi(t = t_0) = \Psi(t_0)$ gives

$$\Psi(t) = \exp\left[-\frac{i(t - t_0)E}{\hbar}\right]\Psi(t_0) \tag{4.14}$$

which agrees with the previous result [equation (4.13)] when $t_0 = 0$. Equation (4.14) can be rewritten as

$$\Psi(t) = \exp\left(-\frac{itE}{\hbar}\right)\exp\left(\frac{it_0E}{\hbar}\right)\Psi(t_0) \tag{4.15a}$$

$$= \exp\left(-\frac{itE}{\hbar}\right)U(t,\, t_0)\,\Psi(t_0) \tag{4.15b}$$

$$= \exp\left(-\frac{itE}{\hbar}\right)\Psi(t_0 - t_0) \tag{4.15c}$$

$$= \exp\left(-\frac{itE}{\hbar}\right)\Psi(0) \tag{4.15d}$$

If some physical event occurred in the system at $t = 0$, then equation (4.13) gives the temporal evolution of the system from $t = 0$, and time does not appear in $\Psi(0)$ on the right side of equation (4.13). Similarly, if t_0 in equation (4.14) is the time at which some event occurred in the system, then equation (4.14) gives the temporal evolution of the system after than event, and $\Psi(t_0)$ is not a function of time variable t. However, if the new time origin one chooses if unrelated to any internal event in the system, the new time origin simply becomes a new independent variable, and separation of the time variable has not been accomplished. We will see below that these same results are obtained with the time displacement operator $\mathcal{T}(\mathbf{R}_0)$ if the new coordinate origin $\mathbf{R}_0$ is just some arbitrary new origin not referenced to the locations of the electrons and nuclei composing the system.

4.3 Spatial translation symmetry

The spatial translation symmetry operator $\mathcal{T}$ is defined by the effect on the eigenfunction $\Psi(\mathbf{R}_1,\, \mathbf{R}_2,\, \ldots,\, \mathbf{R}_N)$ caused by changing the origin of spatial coordinates from $\mathbf{R} = 0$ to some new origin $\mathbf{R}_0(X_0,\, Y_0,\, Z_0)$,

$$\mathcal{T}(\mathbf{R}_1, \mathbf{R}_2, \ldots, \mathbf{R}_N, \mathbf{R}_0)\,\Psi(\mathbf{R}_1, \mathbf{R}_2, \ldots, \mathbf{R}_N) = \Psi(\mathbf{r}_1, \mathbf{r}_2, \ldots, \mathbf{r}_N, \mathbf{R}_0) \tag{4.16}$$

where

$$\mathbf{r}_i = \mathbf{R}_i - \mathbf{R}_0, \qquad i = 1, 2, \ldots, N. \tag{4.17}$$

The finite translation operator $\mathcal{T}$ will be constructed as an infinite sequence of infinitesimal translations. The translation operator $\Delta\mathcal{T}_1$ for the infinitesimal translation along the X axis $\Delta X_1 = X_1 - \Delta X_0$ is given by

$$\Delta\mathcal{T}_1\,\Psi(X_1) = \Psi(X_1 - \Delta X_0) \tag{4.18a}$$

$$\approx 1 - \Delta X_0 \frac{\partial \Psi}{\partial X_1} + \ldots \tag{4.18b}$$

Similar equations hold for the Y_1 and Z_1 components of $\mathbf{R}_1$ and for components of $\mathbf{R}_2, \mathbf{R}_3, \ldots, \mathbf{R}_N$. Spatial derivatives of the eigenfunction are given by

$$\frac{\partial \Psi(X)}{\partial X} = \frac{i}{\hbar}\, P_X \Psi(X) \tag{4.19a}$$

$$\frac{\partial \Psi(Y)}{\partial Y} = \frac{i}{\hbar}\, P_Y \Psi(Y) \tag{4.19b}$$

$$\frac{\partial \Psi(Z)}{\partial Z} = \frac{i}{\hbar}\, P_Z \Psi(Z). \tag{4.19c}$$

Note the sign difference between the right side of equation (4.2) and the above equations. Insertion of equation (4.19a) into equation (4.18b) gives

$$\Delta\mathcal{T}_1\,\Psi(X_1) = 1 - \frac{i}{\hbar}\,\Delta X_0\, P_{X_1} \Psi(X_1) \tag{4.20}$$

The three-dimensional version of the above equation reads

$$\Delta\mathcal{T}_1 = 1 - \frac{i}{\hbar}\Delta\mathbf{R}_0 \cdot \mathbf{P}_1\,\Psi(\mathbf{R}_1) \tag{4.21}$$

The finite translation operator $\mathcal{T}_1$ is the product of infinitesimal translations in which

$$\Delta\mathbf{R}_1 = \lim_{n\to\infty} \frac{\mathbf{R}_0}{n}, \tag{4.22}$$

$$\mathcal{T}_1 = \lim_{n\to\infty} \prod_{k=1}^{n} \left(1 - \frac{i}{\hbar}\frac{\mathbf{R}_0}{n} \cdot \mathbf{P}_1\right) = \lim_{n\to\infty}\left(1 - \frac{i}{\hbar}\frac{\mathbf{R}_0}{n} \cdot \mathbf{P}_1\right)^n \tag{4.23a}$$

$$= \exp\left(-\frac{i}{\hbar}\,\mathbf{R}_0 \cdot \mathbf{P}_1\right) \tag{4.23b}$$

The total translation operator $\mathcal{T}(\mathbf{R}_1, \mathbf{R}_2, \ldots, \mathbf{R}_N, \mathbf{R}_0)$ is the product of the individual translations,

$$\begin{aligned}\mathcal{T}(\mathbf{R}_1, \mathbf{R}_2, \ldots, \mathbf{R}_N, \mathbf{R}_0) &= \mathcal{T}_1(\mathbf{R}_1, \mathbf{R}_0)\,\mathcal{T}_2(\mathbf{R}_2, \mathbf{R}_0)\ldots\mathcal{T}_N(\mathbf{R}_N, \mathbf{R}_0)\\ &= \exp\left(-\frac{i}{\hbar}\mathbf{R}_0\cdot\mathbf{P}_1\right)\exp\left(-\frac{i}{\hbar}\mathbf{R}_0\cdot\mathbf{P}_2\right)\cdots\exp\left(-\frac{i}{\hbar}\mathbf{R}_0\cdot\mathbf{P}_N\right)\end{aligned} \tag{4.24}$$

Because the linear momentum operators for the different electrons and nuclei commute, the product of the individual translation operators can be written in any order. The first two terms of the above product can be written

$$\begin{aligned}\mathcal{T}_1(\mathbf{R}_1, \mathbf{R}_0)\,\mathcal{T}_2(\mathbf{R}_2, \mathbf{R}_0) &= \exp\left(-\frac{i}{\hbar}\mathbf{R}_0\cdot\mathbf{P}_1\right)\exp\left(-\frac{i}{\hbar}\mathbf{R}_0\cdot\mathbf{P}_2\right)\\ &= \exp\left[-\frac{i}{\hbar}\mathbf{R}_0\cdot(\mathbf{P}_1+\mathbf{P}_2)\right]\exp\,[\mathbf{R}_0\cdot\mathbf{P}_1, \mathbf{R}_0\cdot\mathbf{P}_2]\Big/2\end{aligned} \tag{4.25}$$

in which the rightmost term is the commutator of $\mathbf{R}_0\cdot\mathbf{P}_1$ and $\mathbf{R}_0\cdot\mathbf{P}_2$. The new origin $\mathbf{R}_0$ is a fixed vector, not an operator, and it commutes with the linear momentum operators $\mathbf{P}_i$,

$$\exp\,[\mathbf{R}_0\cdot\mathbf{P}_i, \mathbf{R}_0\cdot\mathbf{P}_i]/2 = 1, \tag{4.26}$$

and the final result for the translation operator is

$$\mathcal{T}(\mathbf{R}_1, \mathbf{R}_2, \ldots, \mathbf{R}_N, \mathbf{R}_0) = \exp\left[-\frac{i}{\hbar}\mathbf{R}_0\cdot\mathbf{P}\right] \tag{4.27a}$$

$$= \exp\left[-\frac{i}{\hbar}\mathbf{P}\cdot\mathbf{R}_0\right] \tag{4.27b}$$

in which $\mathbf{P}$ is the total linear momentum operator,

$$\mathbf{P} = \sum_{i=1}^{N}\mathbf{P}_i. \tag{4.28}$$

If the new origin of coordinates $\mathbf{R}_0$ is not referenced to the locations of the mass points from which the atomic or molecular system is composed, the cartesian components of $\mathbf{R}_0(X_0, Y_0, Z_0)$ add three independent variables to the problem. If the new origin is referenced to the internal spatial coordinates of the system, *and* if a two-body reduction is applied to two of the mass points, *and* if the coordinates of the center of mass $\mathbf{R}_{CM}(X_{CM}, Y_{CM}, Z_{CM})$ are introduced as variables of the system, then application of the translation operator introduces no new independent variables and separates the three spatial components of the center of mass. An example of this will be seen below when the translation operator $\mathcal{T}(\mathbf{R}, \mathbf{R}_0)$ is applied to the diatomic (4.1) eigenfunction.

The time translation operator $\mathcal{U}(t, t_0)$ always involves two coordinate systems, the original time axis and the translated time axes. The spatial translation operator $\mathcal{T}(\mathbf{R}, \mathbf{R}_0)$ always involves two Cartesian coordinate systems, the original system whose origin lies at 0 and the translated system whose origin is $\mathbf{R}_0$. The rotation operator $\mathcal{R}$ below will always involve two coordinate systems, the original system and the rotated system. When do coordinates of translated or rotated coordinate systems become coordinates describing physical motion? As seen above, coordinates of the new coordinate system become internal coordinates describing physical motion when the time origin is referenced to an event in the system and the coordinate origin is referenced to locations of the mass points composing the system. Translation of coordinates to new origins having no physical connection to the system under study provides a new mathematical viewpoint of the system, but are excluded as coordinates describing physical motion. When the new origins of coordinate transformation are physically referenced to the system, the invariance of values and equations under symmetry transformations becomes conservation of energy, linear momentum, and (as shown below) angular momentum during physical motion.

4.4 Two-body symmetry

The model for an atomic or molecular system contains a minimum of two mass points. The first step in finding suitable internal coordinates for any such system is selection of two particles for the two-body reduction. This is not historically the first step nor it is the first step in many textbook treatments of atomic and molecular systems, but sooner or later as approximations are dropped in favor of higher degrees of approximation, one will select two mass points located at $\mathbf{R}_1$ and $\mathbf{R}_2$, and replace their motion with that of two fictitious particles one located $\mathbf{r} = \mathbf{R}_1 - \mathbf{R}_2$ and having reduced mass μ,

$$\mu = \frac{m_1 m_2}{m_1 + m_2}, \tag{4.29}$$

and a second fictitious particle of mass $\mathcal{M}$,

$$\mathcal{M} = m_1 + m_2, \tag{4.30}$$

located at the center of mass of the two particles,

$$\mathbf{R}_0 = \frac{m_1 \mathbf{R}_1 + m_2 \mathbf{R}_2}{m_1 + m_2}. \tag{4.31}$$

The change of variables

$$\mathbf{R}_1, \mathbf{R}_2 \rightarrow \mathbf{r}, \mathbf{R}_0 \tag{4.32}$$

in which

$$\mathbf{r} = \mathbf{R}_1 - \mathbf{R}_2 \tag{4.33a}$$

$$= \mathbf{r}_1 - \mathbf{r}_2 \tag{4.33b}$$

and the chain rule, for example,

$$\frac{\partial \Psi}{\partial X_1} = \frac{\partial \Psi}{\partial x}\frac{dx}{dX_1} + \frac{\partial \Psi}{\partial X_0}\frac{dX_0}{dX_1} = \frac{\partial \Psi}{\partial x} + \frac{m_1}{\mathcal{M}}\frac{\partial \Psi}{\partial X_0} \tag{4.34a}$$

$$\frac{\partial \Psi}{\partial X_2} = \frac{\partial \Psi}{\partial x}\frac{dx}{dX_2} + \frac{\partial \Psi}{\partial X_0}\frac{dX_0}{dX_2} = -\frac{\partial \Psi}{\partial x} + \frac{m_2}{\mathcal{M}}\frac{\partial \Psi}{\partial X_0}, \tag{4.34b}$$

give the result that the total linear momentum **P** of the two masses in terms the new variables is the linear momentum of the center of mass of the two masses,

$$\mathbf{P} = \mathbf{P}_1 + \mathbf{P}_2 \tag{4.35a}$$

$$= \mathbf{p} + \frac{m_1}{\mathcal{M}}\mathbf{P}_0 - \mathbf{p} + \frac{m_2}{\mathcal{M}}\mathbf{P}_0 \tag{4.35b}$$

$$= \mathbf{P}_0 \tag{4.35c}$$

with

$$\mathbf{p} = \frac{\hbar}{i}\left[\frac{\partial}{\partial x} + \frac{\partial}{\partial y} + \frac{\partial}{\partial z}\right]. \tag{4.36}$$

The inverse change of variables,

$$\mathbf{R}_1 = \mathbf{R}_0 + \frac{m_2}{\mathcal{M}}\mathbf{r} \tag{4.37a}$$

$$\mathbf{R}_2 = \mathbf{R}_0 - \frac{m_1}{\mathcal{M}}\mathbf{r}, \tag{4.37b}$$

and the chain rule give the internal (relative) linear momentum **p** in terms of the original coordinates $\mathbf{R}_1$ and $\mathbf{R}_2$,

$$\mathbf{p} = \frac{m_2\mathbf{P}_1 - m_1\mathbf{P}_2}{m_1 + m_2}. \tag{4.38}$$

When the system in question contains only two mass points, the new coordinate origin expressed in the original coordinate system $\mathbf{R}_0$ is the center of total mass $\mathbf{R}_{\mathrm{CM}}$. Two-body symmetry separates the linear momentum of the two particles from their relative motion. Separation of the total linear momentum in a two particle system is accomplished when the coordinate transformation

$$\mathbf{R}_1, \mathbf{R}_2 \rightarrow \mathbf{r}, \mathbf{R}_{\mathrm{CM}} \tag{4.39}$$

is performed. Application of the translation operator $\mathcal{T}(\mathbf{R}_1, \mathbf{R}_2, \mathbf{R}_{\mathrm{CM}})$ to the eigenfunction $\Psi(\mathbf{R}_1, \mathbf{R}_2)$ produces this transformation (change of variables).

In a generic system consisting of more than two mass points, application of the two-body reduction still dictates that the new coordinate origin be at the center of mass of two mass points $\mathbf{R}_0$, and $\mathbf{R}_0$ no longer coincides with center of total mass $\mathbf{R}_{\mathrm{CM}}$. In this case the coordinate transformation is

$$\mathbf{R}_1, \mathbf{R}_2, \ldots, \mathbf{R}_N \rightarrow \mathbf{r}_1, \mathbf{r}_2, \ldots, \mathbf{r}_{N-1}, \mathbf{R}_{\mathrm{CM}} \tag{4.40}$$

4.5 Time and spatial translations together

The symmetry operators $\mathcal{U}$ and $\mathcal{T}$ are unitary

$$\mathcal{T}^{\dagger}\mathcal{T}\mathcal{U}^{\dagger}\mathcal{U}\Psi = \Psi. \tag{4.41}$$

Applied to generic eigenfunction $\Psi(\mathbf{R}_1, \mathbf{R}_2, \ldots, \mathbf{R}_N, t)$ in which N is the number of mass points,

$$\begin{aligned}\mathcal{T}^{\dagger}\mathcal{T}\mathcal{U}^{\dagger}\mathcal{U}\Psi(\mathbf{R}_1, \mathbf{R}_2, \ldots, \mathbf{R}_N, t) &= \mathcal{T}^{\dagger}\mathcal{U}^{\dagger}\Phi(\mathbf{r}_1, \mathbf{r}_2, \ldots, \mathbf{r}_{N-2}, \mathbf{r}) \\ &= \exp\left(\frac{i}{\hbar}\mathbf{R}_0 \cdot \mathbf{P}\right)\exp\left[-\frac{i}{\hbar}t\, E\right] \\ &\quad \times \Phi(\mathbf{r}_1, \mathbf{r}_2, \ldots, \mathbf{r}_{N-2}, \mathbf{r}).\end{aligned} \tag{4.42}$$

A superposition of the exponentials

$$\Psi_{\mathrm{CM}}(\mathbf{R}_0, t) = \exp\left[i\left(\frac{1}{\hbar}\mathbf{R}_0 \cdot \mathbf{P} - \omega t\right)\right] \tag{4.43}$$

with $\omega = E/\hbar$ yielding a wave packet describing linear motion the molecule as a whole.

4.6 Rotational symmetry

The eigenfunction $\Phi(\mathbf{r}_1, \mathbf{r}_2, \ldots, \mathbf{r}_{N-2}, \mathbf{r})$ is expressed in generic coordinates and quantum numbers. Rewritten for the diatomic molecule,

$$\langle \mathbf{r}_1, \mathbf{r}_2, \ldots, \mathbf{r}_N, \mathbf{r}\, |nvJM\rangle = \langle \rho, \zeta, \chi, \mathbf{r}_2, \ldots, \mathbf{r}_N, r, \theta, \phi\, |nvJM\rangle, \tag{4.44}$$

this eigenfunction is identical to that presented in the framework for the Wigner–Witmer eigenfunction (WWE), see equation (3.11). Although the Euler angles α, β, and γ and the rotation operator $\mathcal{R}(\alpha, \beta, \gamma)$ are not properly introduced until chapter 8, but they have already been used in chapter 3, see equations (3.6)–(3.16). Analogous to the presentation of the framework for WWE, the Wigner–Witmer diatomic eigenfunction for the diatomic molecule is written as

$$\langle \rho, \zeta, \chi, \mathbf{r}_2, \ldots, \mathbf{r}_N, r, \theta, \phi\, |nvJM\rangle = \sum_{\Omega=-J}^{J} \left\langle \rho, \zeta, \mathbf{r}'_2, \ldots, \mathbf{r}'_N, r\, |nv\right\rangle D^{J^*}_{M\Omega}(\phi, \theta, \chi). \tag{4.45}$$

That the Euler angles in the Wigner–Witmer diatomic eigenfunction (4.45) are both angles of coordinate rotation and angles of mechanical rotation is extremely atypical. Normally, one must maintain a clear distinction between coordinate (passive) rotation and mechanical (active) rotation. The first two Euler angles ϕ and θ are also exactly separable in the hydrogen atom eigenfunction and serve as both angles of coordinate rotation and mechanical rotation. The angles ϕ and θ in the hydrogen atom are identical to the same angles in the diatomic molecule. In both cases, θ is the polar angle and ϕ is the azimuthal angle of the vector $\mathbf{r}$ drawn from the origin taken to be the center of mass of two particles, the electron and proton of the hydrogen atom and the two nuclei in the diatomic molecule.

References

[1] Wigner E and Witmer E E 1928 *Z. Phys.* **51** 859
Hettema H 2000 *Quantum Chemistry: Classic Scientific Papers* (Singapore: World Scientific) p 287
[2] Weissbluth M 1978 *Atoms and Molecules* (New York: Academic)

IOP Publishing

Quantum Mechanics of the Diatomic Molecule with Applications

Christian G Parigger and James O Hornkohl

Chapter 5

Diatomic formula inferred from the Wigner–Witmer eigenfunction

In this chapter, a formula useful in diatomic spectroscopy is derived from the Wigner–Witmer eigenfunction [1] (equation (3.16)).

Equation (3.2d) holds in a coordinate system, and equation (3.3b) holds in a second coordinate system obtained from the first by coordinate rotation, but on the right side of equation (3.16) the only thing $J'_{\pm}$ can operate on is $D^{J^*}_{M\Omega}(\phi, \theta, \chi)$. Equation (3.3b) shows what happens when $J'_{\pm}$ is applied to $|J\Omega\rangle$, but what happens when $J'_{\pm}$ is applied to $D^{J^*}_{M\Omega}(\phi, \theta, \chi)$? This mathematical question is answered in the following paragraphs.

Angular momentum theory gives the result of applying the raising and lowering operators to the left side of the Wigner–Witmer equation (3.16), and from this, we can find how $J_{\pm}$ operates on $D^{J^*}_{M\Omega}(\phi, \theta, \chi)$,

$$\begin{aligned}
&\langle \rho, \zeta, \chi, \mathbf{r}_2, \ldots, \mathbf{r}_N, r, \theta, \phi\, |J_{\pm}|nvJM\rangle \\
&= \sqrt{J(J+1) - M(M \pm 1)}\,\langle \rho, \zeta, \chi, \mathbf{r}_2, \ldots, \mathbf{r}_N, r, \theta, \phi\, |nvJ, M \pm 1\rangle \\
&= \sqrt{J(J+1) - M(M \pm 1)} \sum_{\Omega=-J}^{J} \langle \rho, \zeta, \mathbf{r}'_2, \ldots, \mathbf{r}'r\, |nv\rangle D^{J^*}_{M\pm 1,\Omega}(\phi, \theta, \chi) \\
&= \sum_{\Omega=-J}^{J} \langle \rho, \zeta, \mathbf{r}'_2, \ldots, \mathbf{r}'r\, |nv\rangle J_{\pm}\, D^{J^*}_{M\Omega}(\phi, \theta, \chi).
\end{aligned} \tag{5.1}$$

The eigenfunctions $\langle \rho, \zeta, \mathbf{r}'_2, \ldots, \mathbf{r}'r\, |nv\rangle$ are independent, and the result

$$J_{\pm} D^{J^*}_{M\Omega}(\phi, \theta, \chi) = \sqrt{J(J+1) - M(M \pm 1)}\; D^{J^*}_{M\pm 1,\Omega}(\phi, \theta, \chi) \tag{5.2}$$

holds for each term in the above summations and is intuitively pleasing. The raising and lowering operators raise and lower M on $D^{J^*}_{M\pm1,\Omega}(\phi, \theta, \chi)$. However, on taking the complex conjugate of both side of the above equation [note from (3.2c) that $J^*_{\pm} = J_{\mp}$],

$$J_{\mp}D^J_{M\Omega}(\phi, \theta, \chi) = \sqrt{J(J+1) - M(M \pm 1)}\; D^J_{M\pm1,\Omega}(\phi, \theta, \chi), \tag{5.3}$$

and intuition is now faring less well. The raising and lowering operators lower and raise M, respectively, when applied to $D^J_{M\mp1,\Omega}(\phi, \theta, \chi)$. The above is a little more easily read if one swaps all the $\pm$ with $\mp$ and vice versa,

$$J_{\pm}D^J_{M\Omega}(\phi, \theta, \chi) = \sqrt{J(J+1) - M(M \mp 1)}\; D^J_{M\mp1,\Omega}(\phi, \theta, \chi). \tag{5.4}$$

The above describes operation $J_{\pm}$ on the D-functions and its complex conjugate, but we have missed the goal of learning how the rotated raising and lowering operators $J'_{\pm}$ operate on the D-function. To accomplish this goal, we need to turn the Wigner–Witmer equation (3.16) inside out, to express the rotated eigenfunction as a sum of un-rotated eigenfunctions.

The rotation matrix elements form a $(2J+1) \times (2J+1)$ unitary matrix. Multiplying equation (3.12) by $D^J_{M'\Omega}(\alpha, \beta, \gamma)$ and summing over M' one obtains

$$\begin{aligned}&\langle \rho, \zeta, \chi', \mathbf{r}'_2, \ldots, \mathbf{r}'_N, r, \theta', \phi' \,|nvJ\Omega\rangle \\ &\quad = \sum_{M=-J}^{J} \langle \rho, \zeta, \chi, \mathbf{r}_2, \ldots, \mathbf{r}_N, r, \theta, \phi \,|nvJM\rangle\, D^J_{M\Omega}(\alpha, \beta, \gamma).\end{aligned} \tag{5.5}$$

Note that when equation (3.12) was inverted to yield the above, the Euler angles of mechanical rotation and the Euler angles of coordinate rotation were unrelated. Having inverted equation (3.12) we choose Euler angles of coordinate rotation to remove the angular dependence from $\langle \rho, \zeta, \chi, \mathbf{r}_2, \ldots, \mathbf{r}_N, r, \theta, \phi \,|nvJM\rangle$,

$$\phi' = \phi + \alpha, \tag{5.6a}$$

$$\theta' = \theta + \beta, \tag{5.6b}$$

$$\chi' = \chi + \gamma. \tag{5.6c}$$

Note that the α, β, and γ chosen in equations (5.6) are the negative to those in equations (3.13). The inverse of the rotation operator $\mathcal{R}^{-1}(\alpha, \beta, \gamma)$ is $\mathcal{R}(-\gamma, -\beta, -\alpha)$ [2]. The choices in equations (5.6) yield

$$\begin{aligned}&\langle \rho, \zeta, \chi', \mathbf{r}'_2, \ldots, \mathbf{r}'_N, r, \theta', \phi' \,|nvJ\Omega\rangle \\ &\quad = \sum_{M=-J}^{J} \langle \rho, \zeta, \mathbf{r}_2, \ldots, \mathbf{r}_N, r \,|nv\rangle\, D^J_{M\Omega}(\phi', \theta', \chi')\end{aligned} \tag{5.7}$$

We can now apply $J'_{\pm}$ on both sides of this equation, and using the standard result (3.3b) mathematically determine the operation of $J'_{\pm}$ on $D^J_{M\Omega}(\phi', \theta', \chi')$. Operation of $J'_{\pm}$ to both sides of the above equation gives

$$\begin{aligned}&\sqrt{J(J+1)-\Omega(\Omega\pm 1)}\,\langle \rho, \zeta, \chi', \mathbf{r}'_2, \ldots, \mathbf{r}'_N, r, \theta', \phi' \,|nvJ, \Omega \pm 1\rangle \\ &\quad \sqrt{J(J+1)-\Omega(\Omega\pm 1)} \sum_{M=-J}^{J} \langle \rho, \zeta, \mathbf{r}_2, \ldots, \mathbf{r}_N, r \,|nv\rangle\, D^J_{M,\Omega\pm 1}(\phi', \theta', \chi') \\ &\quad = \sum_{M=-J}^{J} \langle \rho, \zeta, \mathbf{r}_2, \ldots, \mathbf{r}_N, r \,|nvJM\rangle\, J'_{\pm}\, D^J_{M\Omega}(\phi', \theta', \chi')\end{aligned} \tag{5.8}$$

from which, after dropping primes on the Euler angles, one obtains

$$J'_{\pm}\, D^J_{M\Omega}(\phi, \theta, \chi) = \sqrt{J(J+1)-\Omega(\Omega\pm 1)}\, D^J_{M,\Omega\pm 1}(\phi, \theta, \chi). \tag{5.9}$$

This result, like equation (5.2), is intuitively acceptable. The rotated raising and lowering operators $J'_{\pm}$ raise and lower Ω on $D^J_{M\Omega}(\phi, \theta, \chi)$. On taking the complex conjugate of both sides of the above equation, one finds a less intuitively pleasing result,

$$J'_{\pm}\, D^{J^*}_{M\Omega}(\phi, \theta, \chi) = \sqrt{J(J+1)-\Omega(\Omega\mp 1)}\, D^{J^*}_{M,\Omega\mp 1}(\phi, \theta, \chi). \tag{5.10}$$

This equation is a mathematical result, not an accepted 'phase convention'. The mysterious difference between equation (3.3b) and equation (5.10) disappears when one understands that $|J\Omega\rangle$ is an eigenstate of angular momentum but neither $D^{J^*}_{M\Omega}(\phi, \theta, \chi)$ or $D^J_{M\Omega}(\phi, \theta, \chi)$ is.

The effects of raising and lowering operators are further discussed in appendix B. The importance of equation (5.10) derives from its use in calculation of Hund's case (a) matrix representation of the diatomic Hamiltonian.

References

[1] Wigner E and Witmer E E 1928 *Z. Phys.* **51** 859
Hettema H 2000 *Quantum Chemistry: Classic Scientific Papers* (Singapore: World Scientific) p 287

[2] Arfken G B, Weber H J and Harris F E 2012 *Mathematical Methods for Physicists, A comprehensive Guide* 7th edn (New York: Academic)

IOP Publishing

Quantum Mechanics of the Diatomic Molecule with Applications

Christian G Parigger and James O Hornkohl

Chapter 6

Hund's cases (a) and (b)

6.1 Introduction

In spectroscopic notation, the total angular momentum **F** of the diatomic molecule is given by

$$\mathbf{F} = \mathbf{L} + \mathbf{R} + \mathbf{S} + \mathbf{T} \tag{6.1a}$$

$$= \mathbf{J} + \mathbf{T} \tag{6.1b}$$

$$\mathbf{J} = \mathbf{L} + \mathbf{R} + \mathbf{S} \tag{6.1c}$$

$$= \mathbf{N} + \mathbf{S} \tag{6.1d}$$

$$\mathbf{N} = \mathbf{L} + \mathbf{R} \tag{6.1e}$$

$$\mathbf{T} = \mathbf{I}_{\mathrm{a}} + \mathbf{I}_{\mathrm{b}} \tag{6.1f}$$

in which $\mathbf{I}_{\mathrm{a}}$ and $\mathbf{I}_{\mathrm{b}}$ are the nuclear spins and **T** is their sum (the a on $\mathbf{I}_{\mathrm{a}}$ and b on $\mathbf{I}_{\mathrm{b}}$ are the labels for the two nuclei and are unrelated to Hund's cases (a) and (b) angular momentum coupling models). The total orbital angular momentum **N** is the sum **L** of the individual electronic orbital angular momenta ℓ_i,

$$\mathbf{L} = \sum_{i=1}^{N} \ell_i, \tag{6.2}$$

plus the orbital angular momentum **R** of the two nuclei. The total electronic spin is the sum of the individual electronic spins,

doi:10.1088/978-0-7503-1989-8ch6

$$\mathbf{S} = \sum_{i=1} \mathbf{s}_i. \tag{6.3}$$

In spectroscopic notation, **J** is the total angular momentum including everything except nuclear spin. In this chapter, nuclear spin is ignored, and the spectroscopic J, M, and Ω quantum numbers are treated as if they represent the true total angular momentum.

Orbital and spin momenta produce magnetic moments. It is obvious from equations(6.1) that even when nuclear spin is ignored there are many ways in which magnetic moments produced by **L**, **R**, their sum $\mathbf{N} = \mathbf{L} + \mathbf{R}$, and **S** could interact. In Hund's angular momentum coupling models, a particular interaction is assumed to dominate all others. Only Hund's cases (a) and (b) are considered in this book, and they will be used as mathematically complete bases, not mechanical models. The following shows how cases (a) and (b) basis functions are obtained from the Wigner–Witmer eigenfunction.

6.2 Case (b) basis functions

Construction of case (b) basis functions will be familiar to those familiar with ℓs coupling in atoms where one builds $|jm\rangle$ states from orbital angular momentum states $|\ell m\rangle$ and spin states $|sm_s\rangle$ states. In the diatomic case, one builds $|JM\rangle$ states from total orbital angular momentum states $|NM_N\rangle$ and, ignoring nuclear spin, total electronic spin states $|SM_S\rangle$.

In the absence of both electronic and nuclear spin, the Wigner–Witmer eigenfunction is

$$\begin{aligned}\langle \mathbf{r}_1, \mathbf{r}_2, \ldots, \mathbf{r}_{\mathcal{N}}, \mathbf{r}\, |nvNM_N\rangle = & \sum_{\Lambda=-N}^{N} \langle \rho, \zeta, \mathbf{r}_2, \ldots, \mathbf{r}_{\mathcal{N}}, r\, |nv\rangle \\ & \times D^{N*}_{M_N\Lambda}(\phi, \theta, \chi).\end{aligned} \tag{6.4}$$

We have used $\mathcal{N}$ as the number of electrons to avoid confusion with the quantum number N. Dropping the summation over Λ from the right side and renormalization gives the function

$$\begin{aligned}\langle \rho, \zeta, \chi, \mathbf{r}'_2, \ldots, \mathbf{r}'_{\mathcal{N}}, r, \theta, \phi\, |nvNM_N\Lambda\rangle = & \sqrt{\frac{2N+1}{8\pi^2}} \langle \rho, \zeta, \mathbf{r}_2, \ldots, \mathbf{r}_{\mathcal{N}}, r\, |nv\rangle \\ & \times D^{N*}_{M_N\Lambda}(\phi, \theta, \chi).\end{aligned} \tag{6.5}$$

In case (b), the mechanical coupling of the magnetic moment of the spin state $|SM_S\rangle$ to the magnetic moment of the orbital $|NM_N\rangle$ state is assumed to be negligibly small. This is ideally suited to building total angular momentum states $|JM\rangle$ using Clebsch–Gordan coefficients $\langle NM_N;\, SM_S\, |JM\rangle$,

$$|JM\rangle = \sum_{M_N=-N}^{N} \sum_{M_S=-S}^{S} |NM_N\rangle \otimes |SM_S\rangle \langle NM_N;\ SM_S\ |JM\rangle. \tag{6.6}$$

An unfortunate notational convention leaves use of the angular momentum coupling coefficients most valid when the mechanical coupling is negligible. Other notational features of the above equation deserve explanation. The tensor (or direct or Kronecker) product $|NM_N\rangle \otimes |SM_S\rangle$ is not written as $|NM_N\rangle\ |SM_S\rangle$ because the latter lacks mathematical meaning, e.g. $|NM_N\rangle$ and $|SM_S\rangle$ are normally regarded as column matrices having $(2N + 1)$ rows and $(2S + 1)$ rows, and their product is not commensurate with standard matrix multiplication[1, 2]. Also, a semicolon appears in the Clebsch–Gordan coefficient $\langle NM_N;\ SM_S|JM\rangle$ to remind one that it is merely a number, not a collection bras and kets[1].

The case (b) basis function is obtained when equation (6.5) is inserted into equation (6.6),

$$\begin{aligned} |b\rangle &= \langle \rho, \zeta, \chi, \mathbf{r}_2, \dots, \mathbf{r}_N, r, \theta, \phi\ |nvN\Lambda S\rangle \\ &= \sqrt{\frac{2N+1}{8\pi^2}} \langle \rho, \zeta, \mathbf{r}_2, \dots, \mathbf{r}_N, r\ |nv\rangle \\ &\times \sum_{M_N=-N}^{N} \sum_{M_S=-S}^{S} \langle NM_N;\ SM_S\ |SM_S\rangle |SM_S\rangle\ D^{N*}_{M_N\Lambda}(\phi, \theta, \chi). \end{aligned} \tag{6.7}$$

The direct product symbol $\otimes$ should appear between $|SM_S\rangle$ and $D^{N*}_{M_N\Lambda}(\phi, \theta, \chi)$ in the case (b) basis function, but in practice one never attempts to form the direct product of $|SM_S\rangle$ with the Wigner D-function and the symbol $\otimes$ is conventionally dropped.

It is important that one remember that the summation over Λ in equation (6.4) has been dropped. As a result, the case (b) basis functions are not *energy* eigenfunctions, not *angular momentum* eigenfunctions, and not *parity* eigenfunctions. For example, a sum of case (b) basis functions,

$$|JM\rangle = \sum_{\Lambda=-N}^{N} |b\rangle \langle b\ |JM\rangle, \tag{6.8}$$

is required to build the angular momentum eigenfunction. The case (b) basis possesses three magnetic quantum numbers, M_N, Λ, and M_S, two more than possessed by a quantum mechanical state of angular momentum.

Hund's case (b) basis is mathematically complete. In principle, one can successfully perform in the case (b) basis any calculation that can be performed in any other basis or with the exact eigenfunction.

[1] Notations for the Clebsch–Gordon coefficients are so varied that one is almost at liberty to invent new ones. For example, coefficients like Zare's [3] help one to avoid double subscripts and subscripted superscripts.

6.3 Case (a) eigenfunctions

When the summation over Ω is dropped from the right side of the Wigner–Witmer eigenfunction (3.16) and the result is renormalized, one obtains the function

$$\begin{aligned}&\langle \rho, \zeta, \chi, \mathbf{r}'_2, \ldots, \mathbf{r}'_N, r, \theta, \phi \,|nvJM\Omega\rangle \\ &= \sqrt{\frac{2J+1}{8\pi^2}} \langle \rho, \zeta, \mathbf{r}'_2, \ldots, \mathbf{r}'_N, r \,|nv\rangle\, D^{J^*}_{M\Omega}(\phi, \theta, \chi).\end{aligned} \tag{6.9}$$

In case (b), the interaction of the orbital states $|NM_N\rangle$ and $|SM_S\rangle$ is assumed negligibly small, and the $|JM\rangle$ state is built from Clebsch–Gordan coefficients. In case (a), one assumes the electronic spin state $|S\Sigma\rangle$ referenced to rotated coordinates is locked to the orbital state $|N\Lambda\rangle$ also referenced to rotated coordinates. The perfect, complete coupling of $|S\Sigma\rangle$ to $|N\Lambda\rangle$ negates use of Clebsch–Gordon coefficients. The quantum numbers Ω, Λ, and Σ each refer to a z' (internuclear) component of angular momentum. This means that

$$J_{z'} = L_{z'} + S_{z'}, \tag{6.10}$$

$$\Omega = \Lambda + \Sigma. \tag{6.11}$$

Because $\mathbf{J} = \mathbf{L} + \mathbf{S}$, the quantum numbers Λ and Σ are already implied in equation (6.9), but the spin state $|S\Sigma\rangle$ is missing. The case (a) basis function is obtained when the spin ket $|S\Sigma\rangle$ is simply inserted into equation (6.9),

$$\begin{aligned}|a\rangle &= \langle \rho, \zeta, \chi, \mathbf{r}'_2, \ldots, \mathbf{r}'_N, r, \theta, \phi \,|nvJMS\Lambda\Sigma\Omega\rangle \\ &= \sqrt{\frac{2J+1}{8\pi^2}} \langle \rho, \zeta, \mathbf{r}'_2, \ldots, \mathbf{r}'_N, r \,|nv\rangle\, |S\Sigma\rangle\, D^{J^*}_{M\Omega}(\phi, \theta, \chi).\end{aligned} \tag{6.12}$$

The case (a) basis function is one of the more useful mathematical tools of diatomic theory. Using semi-empirical molecular parameters and case (a) basis functions, one can compute the case (a) matrix representation of the Hamiltonian which is then numerically diagonalized. If this is done for the upper and lower states of a band system, the vacuum wavenumbers $\tilde{\nu}$ are the differences between the upper and lower energy eigenvalues, and the Hönl–London line-strength factors $S(J', J)$ are computed from the orthogonal matrices that diagonalized the upper and lower Hamiltonians and the case (a) matrix elements of the transition operator (e.g. the electric dipole operator). A non-vanishing $S(J', J)$ identifies an allowed spectral line.

Like the case (b) basis, the case (a) basis functions are not eigenfunctions of energy, angular momentum, or parity. The complex conjugate of the Wigner D-function has some properties of a $|JM\rangle$ state. For example,

$$\mathbf{J}^2 D^{J^*}_{M\Omega}(\phi, \theta, \chi) = J(J+1)\, D^{J^*}_{M\Omega}(\phi, \theta, \chi), \tag{6.13}$$

$$J_z D^{J^*}_{M\Omega}(\phi, \theta, \chi) = M D^{J^*}_{M\Omega}(\phi, \theta, \chi), \tag{6.14}$$

but then the matter becomes complicated when $J_{z'}$ operates on $D^{J^*}_{M\Omega}(\phi, \theta, \chi)$,

$$J_{z'}D^{J^*}_{M\Omega}(\phi, \theta, \chi) = \Omega D^{J^*}_{M\Omega}(\phi, \theta, \chi). \tag{6.15}$$

Now, there are two magnetic quantum numbers. $|JM\rangle$ and $|J\Omega\rangle$ might possibly represent states of angular momentum, $|JM\Omega\rangle$ cannot. Being mathematically precise about what is or is not an angular momentum eigenfunction is not wasted effort. For example, if one treats the Wigner D function as an angular momentum eigenfunction, operation of $J'_{\pm}$ on it produces a train wreck,

$$J'_{\pm}D^{J^*}_{M\Omega}(\phi, \theta, \chi) = \sqrt{J(J+1) - \Omega(\Omega \mp 1)}\; D^{J^*}_{M,\,\Omega\mp1}(\alpha, \beta, \gamma), \tag{6.16}$$

in which raising/lowering operators whose standard behavior is

$$J'_{\pm}\,|J\Omega\rangle = \sqrt{J(J+1) - \Omega(\Omega \pm 1)}\;|J, \Omega \pm 1\rangle \tag{6.17}$$

become when operating on $D^{J^*}_{M\Omega}(\phi, \theta, \chi)$ lowering/raising operators. If one accepts that a function carrying two magnetic quantum numbers cannot represent an angular momentum state, then the mystery concerning equation (6.16) disappears. Equation (6.16) is simply a mathematical result, not a conundrum in diatomic theory. Equation (6.16) is extensively used in the calculation of case (a) matrix elements. There is also clearly a $N'_{\pm}\, D^{N^*}_{M_N\Lambda}(\phi, \theta, \chi)$ version of this result which is used in case (b) calculations. The importance of equation (6.16) requires that its derivation be given in the following section.

Like the case (b) basis, the case (a) basis is complete. In principle, any calculation can be successfully carried out in the case (a) basis. Typically, calculations in the case (a) basis are analytically less messy than in the case (b) basis.

References

[1] Zettili N 2009 *Quantum Mechanics* 2nd edn (Chichester: Wiley)
[2] Cohen-Tannoudji C, Diu B and Laloë F 1997 *Quantum Mechanics* (New York: Wiley)
[3] Zare R N 1988 *Angular Momentum* (New York: Wiley)

IOP Publishing

Christian G Parigger and James O Hornkohl

Chapter 7

Basis set for the diatomic molecule

This chapter uses Hund's case (a) representation, equation (6.12) of the previous chapter, as the mathematical model of the diatomic molecule. A pictorial model of Hund's case (a) is often presented, e.g. [1], figure 97, or see Bransden and Joachain [2].

The Born–Oppenheimer approximation gives an approximate separation of the diatomic eigenfunction into electronic, vibrational, and rotational parts,

$$\Psi \approx \psi_{\mathrm{el}}\psi_{\mathrm{vib}}\psi_{\mathrm{rot}}. \tag{7.1}$$

In the Wigner–Witmer diatomic eigenfunction, the coordinates ϕ, θ, and χ are exactly separated from the electronic–vibrational coordinates,

$$\begin{aligned} \langle \rho, \zeta, \chi, \mathbf{r}_2, \ldots, \mathbf{r}_N, r, \theta, \phi \,|nvJM\rangle &= \sum_{\Omega=-J}^{J} \langle \rho, \zeta, \mathbf{r}'_2, \ldots, \mathbf{r}'_N, r \,|nv\rangle D^{J^*}_{M\Omega}(\alpha, \beta, \gamma) \\ &= \sum_{\Omega=-J}^{J} (\langle \rho, \zeta, \mathbf{r}'_2, \ldots, \mathbf{r}'_N, r \,|nv\rangle)\langle \phi, \theta, \chi \,|JM\Omega\rangle. \end{aligned} \tag{7.2}$$

The mathematical function $\langle \rho, \zeta, \mathbf{r}'_2, \ldots, \mathbf{r}'_N, r \,|nv\rangle$ is the *electronic–vibrational basis function*. The mathematical function $\langle \phi, \theta, \chi \,|JM\Omega\rangle$,

$$\langle \phi, \theta, \chi \,|JM\Omega\rangle \equiv D^{J^*}_{M\Omega}(\alpha, \beta, \gamma), \tag{7.3}$$

is the *angular momentum basis function*. Calculation of the diatomic eigenfunction requires that one sum the product of electronic–vibrational and angular momentum

doi:10.1088/978-0-7503-1989-8ch7

Table 7.1. Hund's case (a) matrix representation of a hypothetical diatomic Hamiltonian for $J = 2$ and $S = 1$.

$H_{-1,-1}$	$H_{-1,0}$	$H_{-1,1}$	0	0	0	0	0
$H_{0,-1}$	$H_{0,0}$	$H_{0,1}$	0	0	0	0	0
$H_{1,-1}$	$H_{1,0}$	$H_{1,1}$	0	0	0	0	0
0	0	0	$H_{-2,-2}$	$H_{-2,-1}$	$H_{-2,0}$	$H_{-2,1}$	$H_{-2,2}$
0	0	0	$H_{-1,-2}$	$H_{-1,-1}$	$H_{-1,0}$	$H_{-1,1}$	$H_{-1,2}$
0	0	0	$H_{0,-2}$	$H_{0,-1}$	$H_{0,0}$	$H_{0,1}$	$H_{0,2}$
0	0	0	$H_{1,-2}$	$H_{1,-1}$	$H_{1,0}$	$H_{1,1}$	$H_{1,2}$
0	0	0	$H_{2,-2}$	$H_{2,-1}$	$H_{2,0}$	$H_{2,1}$	$H_{2,2}$

basis functions. Equation (7.2) is obviously more complicated than equation (7.1) but the latter is exact, and also more specific. The precise meaning of the word rotational is sometimes unclear. Is $\psi_{\rm rot}$ the angular momentum eigenfunction of the diatomic molecule? The angle χ is both an electronic and an angular momentum variable. Is χ an argument of $\psi_{\rm el}$ or $\psi_{\rm rot}$? Similarly, is the total electronic spin $\mathbf{S}$ associated with $\psi_{\rm el}$ or $\psi_{\rm rot}$?

One knows from first principles of angular momentum theory that for a conservative system the Hamiltonian will be a $(2J + 1) \times (2J + 1)$ matrix. Rotational invariance in isotropic space leads to conservation of the total angular momentum. The quantum numbers J, M, and Ω include spin. However, angular momentum theory does not tell us how the total is composed of its orbital and spin components. One must postulate an angular momentum coupling model that describes how the total $\mathbf{J}$ is formed from, for example, the total orbital angular momentum $\mathbf{N}$ and the total electronic spin $\mathbf{S}$, and, as described in chapter 6, constructs a mathematically complete basis for the $|JM\rangle$ eigenstates of the total angular momentum. If the total electronic spin is to a useful degree of approximation a constant of the motion, (i.e. if S is a so-called 'good quantum number'), the multiplicity $2S + 1$ expands the dimensions of the Hamiltonian to $(2S + 1)(2J + 1) \times (2S + 1)(2J + 1)$. For the commonly occurring values $J = 20$ and $S = 1$, one is contemplating 123×123 matrices. The smallness of the mass of the electron comes to the rescue, and not in exactly the same ways it does in the Born–Oppenheimer approximation. Table 7.1 illustrates hypothetical Hamiltonian for $J = 2$ and $S = 1$.

The angle χ describes rotation of one of the electrons about the internuclear vector at the distance ρ. The Hamiltonian will contain the kinetic energy term

$$T_\chi = -\sum_{\Omega=-J}^{J} \frac{\hbar^2}{2m_{\rm e}\,\rho^2} \frac{\partial^2 \Psi_{nvJ\Omega}}{\partial \chi^2} = \sum_{\Omega=-J}^{J} \frac{\hbar^2 \Omega^2}{2m_{\rm e}\,\rho^2}. \tag{7.4}$$

This result holds if rotation of the nuclei about the internuclear axis is ignored[1].

In the Born–Oppenheimer approximation, the smallness of electronic mass has the nuclei moving much more rapidly than the nuclei. This consideration from classical mechanics leads to the approximate separation of the electronic eigenfunction ψ_e. In the context of the Wigner–Witmer diatomic eigenfunction, the smallness of electronic mass adds a large number of large kinetic energy terms to the Hamiltonian. These large kinetic energy terms come in pairs, the first associated with the basis function $D^{J^*}_{M\Omega}(\alpha, \beta, \gamma)$ and the second with $D^{J^*}_{M,\,-\Omega}(\alpha, \beta, \gamma)$, and from equation (7.4) one sees than the $\langle\phi, \theta, \chi\ |JM\Omega\rangle$ and $\langle\phi, \theta, \chi\ |JM, -\Omega\rangle$ basis states contribute equal amounts of kinetic energy to the Hamiltonian. As a first degree of approximation, we assume that $\langle\phi, \theta, \chi\ |JM\Omega\rangle$ and $\langle\phi, \theta, \chi\ |JM, -\Omega\rangle$ make equal contributions. This turns out be a close approximation.

References

[1] Herzberg G 1950 *Molecular Spectra and Molecular Structure I. Spectra of Diatomic Molecules* 2nd edn (New York: Van Nostrand Reinhold)

[2] Bransden B H and Joachain C J 2003 *Physics of Atom and Molecules* 2nd edn (Harlow: Prentice Hall)

[1] The term $m_e\rho^2$ in equation (7.4) is the moment of inertia of an electron about the internuclear axis. One might falsely assume because nuclear masses are roughly $10^3 - 10^5$ larger that the electronic mass m_e, that the moment of inertia of a nucleus about the internuclear axis is much larger than that of an electron. Actually, the small size of a nucleus overwhelms its large mass, and the moment of inertia of a nucleus about the internuclear axis is much smaller than the electron's. The energy of a state of nuclear rotation about the internuclear axis would be very large, so large in fact that attempting to excite it would destroy the molecule. Thus, rotation of the nuclei about the internuclear axis can be ignored in an existing molecule.

Chapter 8

Quantum theory of angular momentum

8.1 Introduction

The interpretation of atomic and diatomic spectra were early challenges in the development of quantum mechanics and led to the quantum theory of angular momentum. The isotropy of space, rotational invariance, and conservation of the total angular momentum are such general concepts that angular momentum is now omnipresent in physical theory. Angular momentum is treated in quantum mechanics textbooks (e.g. [1–6], and there are several monographs on the topic, i.e. [7–13]. The compendium of [14] is a valuable reference.

In this chapter and throughout this book, a very strict adherence to the mathematical definition of angular momentum is maintained. The mathematical nature of the theory of quantum mechanics can be viewed as a curse or as a valuable tool for recognizing spurious results produced by classical concepts and associated classical intuition. For example, the rotation matrix element $D^{J}_{MM'}(\alpha, \beta, \gamma)$ and its complex conjugate $D^{J*}_{MM'}(\alpha, \beta, \gamma)$ are functions of only angular variables, carry only angular momentum quantum numbers, and are eigenfunctions of a differential equation for the angles. Therefore, the rotation matrix element may be viewed as an angular momentum eigenfunction. The rotation matrix element, written in shorthand as $|JM\Omega\rangle$, is obviously not the eigenstate $|JM\rangle$ or $|J\Omega\rangle$, or in other words, $|JM\Omega\rangle$ is not an eigenstate in a standard description of angular momentum. In the standard description, **J** and only one component of **J**, conventionally chosen to the J_z, describes a quantum mechanical angular momentum state, either $|JM\rangle$ or $|J\Omega\rangle$. In terms of quantum mechanics operators, $\mathbf{J}^2$ and only one component of commute, hence, can be measured simultaneously The mathematical entity $|J\Omega\rangle$ carries two magnetic quantum numbers, therefore, does not satisfy the mathematical properties of angular momentum.

doi:10.1088/978-0-7503-1989-8ch8

The concise distinctions discussed in the previous paragraph are actually essential for diatomic spectroscopy. The formulas for operation of raising J_+ and lowering J_- operators on standard $|JM\rangle$ states are rotationally invariant, like Schrödinger's equation and several other equations encountered in quantum physics,

$$J_{\pm}|JM\rangle = \sqrt{J(J+1) - M(M \pm 1)}\ |J, M \pm 1\rangle, \tag{8.1a}$$

$$J''_{\pm}|J\Omega\rangle = \sqrt{J(J+1) - \Omega(\Omega \pm 1)}\ |J, \Omega \pm 1\rangle, \tag{8.1b}$$

$$\vdots$$

$$J''_{\pm}|JM''\rangle = \sqrt{J(J+1) - M''(M'' \pm 1)}\ |J, M'' \pm 1\rangle. \tag{8.1c}$$

However, what is the origin of the formula of applying angular momentum raising and lowering operators,

$$J'_{\pm} D^{J*}_{M\Omega}(\phi, \theta, \chi) = \sqrt{J(J+1) - \Omega(\Omega \mp 1)}\ D^{J^*}_{M,\Omega \mp 1}(\phi, \theta, \chi), \tag{8.2}$$

frequently applied in diatomic spectroscopy? Equations (8.1) hold for angular momentum states. Equation (8.2), which appears to have a functionality as equation (8.1b), can be correctly viewed as mathematical evidence that the rotation matrix element does not represent an angular momentum state. However, $J'_{\pm}$ operating on $|JM\Omega\rangle = D^{J^*}_{M,\Omega}(\phi, \theta, \chi)$ does not equal $J'_{\pm}$ operating on $|J\Omega\rangle$, because $|JM\Omega\rangle$ cannot be identified with $|J\Omega\rangle$. A sum of $|JM\Omega\rangle$ basis functions, i.e., rotation matrix elements, is required to relate representations of the $|JM\rangle$ and $|J\Omega\rangle$ angular momentum states. In view of diatomic spectroscopy, Equation (8.2) is not a mysterious agreed upon convention [12, 15, 16], but it is rather a mathematical result.

8.2 The standard $|JM\rangle$ angular momentum representation

In addition to equations (8.1) in which

$$J_{\pm} = J_x \pm iJ_y, \tag{8.3}$$

the standard representation provides

$$\mathbf{J}^2|JM\rangle = J(J+1)\,|JM\rangle \tag{8.4}$$

and

$$J_z|JM\rangle = M\ |JM\rangle. \tag{8.5}$$

There are $2J + 1$ values of M,

$$M = -J, -J+1, -J+1, \ldots, J-1, J, \tag{8.6}$$

satisfying equations (8.4) and (8.5). An energy eigenstate $|nJM\rangle$ with n representing quantum numbers required in addition to J and M has a degeneracy of $2J + 1$.

The above results are traditionally derived from the angular momentum commutators,

$$L_xL_y - L_yL_x = iL_z \tag{8.7a}$$

$$L_yL_z - L_zL_y = iL_x, \tag{8.7b}$$

$$L_zL_x - L_xL_z = iL_y, \tag{8.7c}$$

which follow from the classical formula for orbital angular momentum **L** in terms of the linear momentum **p** and distance **r**,

$$\mathbf{L} = \mathbf{r} \times \mathbf{p}, \tag{8.8}$$

and the quantum mechanical linear momentum operators,

$$p_x = \frac{\hbar}{i}\frac{\partial}{\partial x}, \tag{8.9a}$$

$$p_y = \frac{\hbar}{i}\frac{\partial}{\partial y}, \tag{8.9b}$$

$$p_z = \frac{\hbar}{i}\frac{\partial}{\partial z}. \tag{8.9c}$$

Spin cannot be expressed in terms of spatial coordinates but is nevertheless identifiable as angular momentum. With the sum **J** of orbital plus spin momenta replacing **L**, the commutators are often taken to be the quantum mechanical definition of angular momentum,

$$J_xJ_y - J_yJ_x = iJ_z, \tag{8.10a}$$

$$J_yJ_z - J_zJ_y = iJ_x, \tag{8.10b}$$

$$J_zJ_x - J_xJ_z = iJ_y. \tag{8.10c}$$

The operator J_x commutes with its square J_x^2, and **J** therefore commutes with $\mathbf{J}^2$. Although the non-vanishing commutators (8.10) dictate that J_x, J_y, and J_z cannot all be members of the complete set of commuting observables (CSCO), $\mathbf{J}^2$ and one component of **J**, conventionally chosen to be J_z, can be constants of the motion. Again, a mathematical object indicating that two or more components of **J** are constants of motion is not an eigenstate of angular momentum.

8.3 Rotations

The topic of angular momentum is of course closely related to the topic of rotations, in other words, angular momentum can be utilized as generators of rotations. There are two distinct types of rotation, coordinate rotation and mechanical rotation. Invariance is the concept that one can change the details of the mathematical description of a physical state without altering the defining mathematical description of that state. Conservation is the concept that the numerical values of certain variables, e.g. energy, linear momentum, and angular momentum, are constant during mechanical motion. Invariance with respect to transformations implies conservation laws. The homogeneity of time leads to conservation of energy, the homogeneity of space leads to conservation of linear momentum, and the isotropy of space leads to conservation of angular momentum. The concept of angular momentum states of the diatomic molecule is closely related to rotational invariance of the equations that describe the molecule.

Angular momentum $\mathbf{J}(x, y, z)$ is the generator of coordinate rotations α, β, and γ,

$$\mathcal{R}(\alpha, \beta, \gamma) = e^{-i\alpha J_z} e^{-i\beta J_y} e^{-i\gamma J_z}. \tag{8.11}$$

The Hermitian operators J_x, J_y, and J_z in the exponents generate unitary symmetry operators $\mathcal{R}(\alpha)$, $\mathcal{R}(\beta)$, and $\mathcal{R}(\gamma)$. The exponential function that contains the imaginary unit and the angles are as usual interpreted formally as a sum. As shown in quantum mechanics textbooks, for operators A and B,

$$e^A e^B \neq e^{A+B} \tag{8.12}$$

which means that right side of equation (8.11) cannot be written as a single exponential. The standard representations of $|JM\rangle$ and $|J\Omega\rangle$ can be related,

$$\langle \mathbf{r}_1, \mathbf{r}_2, \ldots, \mathbf{r}_N \, |nJM\rangle = \sum_{\Omega=-J}^{J} \langle \mathbf{r}_1, \mathbf{r}_2, \ldots, \mathbf{r}_N \, |nJ\Omega\rangle \, \langle J\Omega \, |JM\rangle, \tag{8.13a}$$

with n denoting other quantum numbers and kept in the equations as needed. Unity in the form $\mathcal{R}(\alpha, \beta, \gamma)\, \mathcal{R}(\alpha, \beta, \gamma)^{-1} = \mathcal{R}(\alpha, \beta, \gamma)\, \mathcal{R}^{\dagger}(\alpha, \beta, \gamma)$ can be applied to both sides,

$$\begin{aligned}
\langle \mathbf{r}_1, \mathbf{r}_2, \ldots, \mathbf{r}_N \, |nJM\rangle &= \sum_{\Omega=-J}^{J} \langle \mathbf{r}_1, \mathbf{r}_2, \ldots, \mathbf{r}_N \, |\mathcal{R}(\alpha, \beta, \gamma) \, |nJ\Omega\rangle \langle J\Omega \, |\mathcal{R}^{\dagger}(\alpha, \beta, \gamma)|JM\rangle \\
&= \sum_{\Omega=-J}^{J} \langle \mathbf{r}'_1, \mathbf{r}'_2, \ldots, \mathbf{r}'_N \, |nJ\Omega\rangle \langle JM \, |\mathcal{R}(\alpha, \beta, \gamma)|J\Omega >^* \qquad (8.13b) \\
&= \sum_{\Omega=-J}^{J} \langle \mathbf{r}'_1, \mathbf{r}'_2, \ldots, \mathbf{r}'_N \, |nJ\Omega\rangle \, D^{J^*}_{M\Omega}(\phi, \theta, \chi).
\end{aligned}$$

The previously discussed equations (8.10) are traditional definition of angular momentum [17]). However, the comparison of the generators of rotations in spherical coordinates,

$$J_\phi = \frac{\hbar}{i} \frac{\partial}{\partial \phi} \tag{8.14a}$$

$$J_\theta = \frac{\hbar}{i} \frac{\partial}{\partial \theta} \tag{8.14b}$$

$$J_\chi = \frac{\hbar}{i}\frac{\partial}{\partial \chi} \tag{8.14c}$$

and of equations (8.9) that describe linear motion indicate that equations (8.14) indeed describe angular motion. It may be preferred to discuss angular momentum utilizing its generators because equations (8.14) appear to associate angular transformation rather than the abstract commutators (8.10).

As one reviews equations (8.14), care must be exercised in the interpretations. Whereas the three commutator equations (8.10) are written in a single coordinate system, each of the three equations (8.14) is written in a different coordinate system. The rotation ϕ is about the z-axis in an xyz system of coordinates. The second rotation θ is about the first intermediate y_1 axis in an Euler rotation of coordinates (see below). The third rotation is about the final z' axis in the full Euler rotation of coordinates. The question arises whether one should work with angular momentum $\mathbf{J}(x, y, z)$ in a single coordinate system, or with $\mathbf{J}(\phi, \theta, \chi)$ that equally reflects the nature of angular momentum.

8.4 Generators of coordinate transformations

The total energy operator H (i.e. the Hamiltonian), the total linear momentum operator $\mathbf{P}$, and the total angular momentum operator $\mathbf{J}$ are the generators of translations in time, translations in space, and rotations in space (see chapter 4). The unitary time translation symmetry operator $\mathcal{U}(t_0)$ gives the time coordinate a new origin, the unitary spatial translation symmetry operator $\mathcal{T}(\mathbf{R}_0)$ gives Cartesian spatial coordinates a new origin, and the unitary rotation symmetry operator $\mathcal{R}(\alpha, \beta, \gamma)$ gives angular coordinates a new origin. In a conservative system, one can always equate the time origin to some physical event. The inverse time translation operator becomes the evolution operator describing the temporal evolution of the system, and the time variable is separated. Invariance during time translation becomes conservation of energy during temporal changes in the system. One can also always reference the new Cartesian coordinate origin to the electrons and nuclei composing the system, and the coordinates of the center of total mass are separated. The inverse of the spatial translation operator becomes the operator describing the total linear motion of the system. Invariance during spatial translation becomes conservation of linear momentum during linear physical motion.

This scheme of using a unitary symmetry operator to separate variables and describe mechanical motion is not applicable for the rotation operator $\mathcal{R}(\alpha, \beta, \gamma)$. The total angular momentum $\mathbf{J}$ can always be defined in terms of the mechanical rotations ϕ, θ, and χ of equations (8.14), and that equations (8.1), (8.4), (8.5), and (8.10) all can be mathematically derived from $\mathbf{J}(\phi, \theta, \chi)$, but almost without exception no real physical systems execute the internal mechanical rotations ϕ, θ, and χ. The total angular momentum is always a constant of the motion for a closed system. Almost never can the total angular momentum for such a system be defined in terms of three exactly separable variables. As shown in the previous chapter, the

Wigner–Witmer diatomic eigenfunction is merely the very general equation for coordinate (passive) rotations written in terms of the internal variables of the diatomic molecule.

The geometry of the diatomic molecule allows one to equate the Euler angles of coordinate rotation to the Euler angles of mechanical rotation thereby removing all angular dependence from the electronic–vibrational eigenfunction $\langle \rho, \zeta, \mathbf{r}'_2, \ldots, \mathbf{r}'_N, r\,|nv\rangle$, and the angles of mechanical rotation are exactly separated. The general equation (3.8) holds only when J, M, and Ω refer to the total angular momentum. Thus, the same quantum numbers in the Wigner–Witmer eigenfunction (3.12) also include all orbital and spin momenta.

References

[1] Schiff L I 1968 *Quantum Mechanics* 3rd edn (London: McGraw-Hill)
[2] Messiah A 1964 *Quantum Mechanics* (Amsterdam: North Holland)
[3] Landau L D and Lifshitz E M 1977 *Quantum Mechanics* 3rd edn (Amsterdam: Butterworth-Heinemann)
[4] Gottfried K and Yan T-M 1998 *Quantum Mechanics: Fundamentals* 2nd edn (New York: Springer)
[5] Zettili N 2009 *Quantum Mechanics* 2nd edn (Chichester: Wiley)
[6] Weinberg S 2015 *Lectures on Quantum Mechanics* 2nd edn (Cambridge: Cambridge University Press)
[7] Rose M E 1995 *Elementary Theory of Angular Momentum* (Mineola, NY: Dover)
[8] Edmonds A R 1974 *Angular Momentum in Quantum Mechanics* 2nd edn (Princeton, NJ: Princeton University Press)
[9] Brink D M and Satchler G R 1993 *Angular Momentum* 3rd edn (Oxford: Oxford University)
[10] Biedenharn L C and Louck J D 2009 *Angular Momentum in Quantum Physics* (Cambridge: Cambridge University Press)
[11] Thompson W J 1994 *Angular Momentum* (New York: Wiley)
[12] Zare R N 1988 *Angular Momentum* (New York: Wiley)
[13] Chaichain M and Hagedorn R 1998 *Symmetries in Quantum Mechanics* (New York: Taylor & Francis)
[14] Varshalovich D A, Moskalev A N and Khersonskii V K 1988 *Quantum Theory of Angular Momentum* (Singapore: World Scientific)
[15] Brown J M and Carrington A 2003 *Rotational Spectroscopy of Diatomic Molecules* (Cambridge: Cambridge University Press)
[16] Lefebvre-Brion H and Field R W 2004 *The Spectra and Dynamics of Diatomic Molecules* (New York: Elsevier/Academic)
[17] Born M and Oppenheimer R 1927 *Ann. Phys.* **84** 457
Hettema H 2000 *Quantum Chemistry: Classic Scientific Papers* (Singapore: World Scientific) p 1

IOP Publishing

Christian G Parigger and James O Hornkohl

Chapter 9

Diatomic parity

How does the parity eigenvalue $p = \pm 1$ vary with the total angular momentum quantum number J? This is a general quantum mechanics question, not a specifically diatomic question. The parity eigenvalue splits into two parts, an angular momentum dependent part and a constant called the intrinsic parity. The following gives the first identification of the intrinsic parity of the diatomic molecule. A Hund basis state for which $\Lambda = 0$ is called a Σ state. The parity operator is the product of an improper rotation that inverts the sign of a single Cartesian component of a coordinate vector and proper rotation that inverts the signs of two Cartesian components. Current diatomic theory places the eigenvalue p_Σ of the one-component parity operator as a superscript on the symbol Σ denoting the $\Lambda = 0$ basis, but currently the one-component eigenvalue is said to have relevance to $\Lambda \neq 0$ basis states. It is seen below that p_Σ is the intrinsic parity of the diatomic molecule, a global value applying to all basis states of the molecule.

The Born–Oppenheimer approximation does not allow one to definitively determine the diatomic parity eigenvalue. The exact separation of the total angular momentum basis in the Wigner–Witmer diatomic eigenfunction is required for exact determination of how the diatomic parity eigenvalue depends on the total angular momentum quantum number J.

The application of diatomic parity to the general diatomic eigenfunction is summarized in recent papers [1, 2] and in appendices E and F. Intrinsic parity is further elaborated in appendix G.

9.1 Parity details

Before showing how the parity of diatomic states follows in a straightforward way from the Wigner–Witmer diatomic eigenfunction, some methods not to be used in the treatment diatomic parity will be itemized.

9.1.1 Parity is rotationally invariant

The parity operator $\mathcal{P}$ changes the signs of Cartesian coordinates,

$$\mathcal{P}\Psi(x, y, z) = \Psi(-x, -y, -z) \tag{9.1}$$

Despite the simple matrix relationship

$$\begin{bmatrix} -1 & 0 & 0 \\ 0 & -1 & 0 \\ 0 & 0 & -1 \end{bmatrix} \begin{bmatrix} x \\ y \\ z \end{bmatrix} = \begin{bmatrix} -x \\ -y \\ -z \end{bmatrix}, \tag{9.2}$$

and the relationship between rotated and fixed coordinates in terms of the direction cosines,

$$\begin{bmatrix} x' \\ y' \\ z' \end{bmatrix} = \begin{bmatrix} c_{11} & c_{12} & c_{13} \\ c_{21} & c_{22} & c_{23} \\ c_{31} & c_{32} & c_{33} \end{bmatrix} \begin{bmatrix} x \\ y \\ z \end{bmatrix}. \tag{9.3}$$

the claim is made that *It is evidently necessary to distinguish clearly between the 'molecule fixed' and the 'laboratory fixed' inversion operation I, since these two operations are not equivalent* [3]. Equations (9.2) and (H.6) are linear transformations. Changing the signs of x, y, and z will obviously change the signs of x', y', and z', and vice versa. Parity is a member of the complete set of commuting observables. The parity operator $\mathcal{P}$, like the other operators of the CSCO, is rotationally invariant. Nevertheless, the idea that inverting signs of fixed coordinates will not invert signs of rotated coordinates has become a standard part of diatomic theory. The reason for this false claim is fairly easy to discern. The discrete inversion of Cartesian signs does not reveal how the parity eigenvalue p varies with the total angular momentum quantum number J. As mentioned elsewhere in this book, angular momentum is angular. How the parity eigenvalue varies with J is not determined by a discrete transformation of Cartesian coordinates but by a discrete transformation of angular coordinates, as shown below.

9.1.2 Spin is immune to the parity operator

Hund's basis contains the spin ket $|S\Sigma\rangle$ referenced to the rotated coordinate system in which the z' axis follows the internuclear vector $\mathbf{r}(r, \theta, \phi)$. The spin ket $|S\Sigma\rangle$ is locked to the internuclear axis. Under coordinate rotation, the spin kets $|SM_S\rangle$ and $|S\Sigma\rangle$ are related by

$$|SM_S\rangle = \sum_{\Sigma=-S}^{S} |S\Sigma\rangle D_{MM'}^{J^*}(\phi, \theta, \chi) \tag{9.4}$$

Spin $\mathbf{S}$ is not expressible in terms of spatial coordinates. The parity operator $\mathcal{P}$ has no defining characteristic other than to alter spatial coordinates. The intrinsic parity, that part of the parity eigenvalue not dependent on J, exists independently of spatial coordinates, but the parity operator does not. Spin is immune to the parity operator. The equation [3–7]

$$\mathcal{P}\,|S\Sigma\rangle = (-)^{S-\Sigma}|S,\,-\Sigma\rangle \tag{9.5}$$

lacks meaning because $|S\Sigma\rangle$ is totally immune to $\mathcal{P}$.

9.1.3 Parity operates on Cartesian coordinates, not angles

The mistake shown in equation (9.5) is the result of thinking that the parity operator can act on Euler angles. The discrete parity transformation acts on Cartesian coordinates. A discrete transformation of the Euler angles can invert the signs of two of the Cartesian coordinates. Thus, the parity operator can be constructed from a discrete transformation of the Euler angles and a one-component parity operator.

9.1.4 Intrinsic parity and Λ doublets

The symbol Σ denotes a diatomic state for which the internuclear component of the total orbital angular momentum **N** is zero (i.e. when the quantum number Λ is zero). A trailing superscript sign is attached, Σ^+ and Σ^-, to denote the eigenvalue p_Σ of a one-dimensional parity operator that inverts the signs of one Cartesian component of the rotated coordinate vectors. The operator $\sigma_v(y'z')$ is used below to invert the signs of the x' components. The eigenvalue p_Σ is the intrinsic parity of the diatomic molecule, a global value. Every state of a given molecule has the intrinsic parity p_Σ. That is, Π states are either Π^+ or Π^-, Δ states are either Δ^+ or Δ^-, and so forth. Every diatomic state for which the internuclear component of the total orbital angular momentum is non-zero is 'Λ doubled'. What this means is that when $\Lambda \neq 0$ each diatomic state comes in two versions, the positive parity version and the negative parity version. This follows directly from the Wigner–Witmer eigenfunction. In current diatomic theory, the value of p_Σ, a value fixed for every state in a given molecule, is considered not applicable when $\Lambda \neq 0$.

9.1.5 Summary of parity details

How the parity eigenvalue $p = \pm 1$ varies with J will be derived below. In this derivation, inversion of the signs of fixed coordinates will be accomplished by inverting the signs of the rotated coordinates, the parity operator will not operate on spin, the parity operator will not operate on the Euler angles, inversion of the signs of two rotated Cartesian coordinates will be accomplished with a discrete transformation of the Euler angles, and intrinsic parity and Λ doublets are treated as separate topics.

9.2 Parity designation

In current practice a diatomic line list typically includes the vacuum wavenumber $\tilde{\nu}$, the upper J' and lower J quantum numbers, and the upper and lower *e*/*f* parity designations. The upper and lower parities are input data to currently used line position fitting programs. In this book, diatomic parity is optional program output, a computed quantity, not required input data. The same eigenvectors that diagonalize the Hamiltonian matrix will also diagonalize the parity matrix. Computation of the latter requires the value p_Σ, the intrinsic parity of the diatomic molecule, and if the molecule is homonuclear the value of gerade/ungerade p_{gu} symmetry of the molecule.

The conversion of parity from required program input data to an optional program output is a major departure from current practice.

9.3 The parity operator

A mathematical function like $\cos(x)$ is called even because $\cos(-x) = \cos(x)$ and another like the $\sin(x)$ is called odd because $\sin(-x) = -\sin(x)$. A function like $f(x) = \cos(x) + \sin(x)$ are neither even nor odd, but its even and odd parts can be found from

$$F(x)_{\text{even}} = \frac{1}{2}[f(x) + f(-x)], \tag{9.6a}$$

$$F(x)_{\text{odd}} = \frac{1}{2}[f(x) - f(-x)]. \tag{9.6b}$$

The parity of a quantum mechanical state is the evenness or oddness of its eigenfunction.

The equation (3.16) for the Wigner–Eckart diatomic eigenfunction uses fixed coordinates on the left and rotated coordinates on the right. The following is concerned with construction of a parity operator that can be applied to either side of the equation. We will find

$$\mathcal{P}_{\text{left}} = \sigma_v(yz)C_2(x) \tag{9.7a}$$

$$\mathcal{P}_{\text{right}} = \sigma_v(y'z')C_2(x'). \tag{9.7b}$$

Note that the above equations are rotationally invariant (i.e. the same equation).

The parity operator $\mathcal{P}$ inverts the signs of each component of all Cartesian coordinates in the Hamiltonian and its eigenfunction. For example, $\mathcal{P}$ can be defined by its operation on the generic eigenfunction $\langle \mathbf{r}_1, \mathbf{r}_2, \ldots, \mathbf{r}_N | nJM\rangle$,

$$\langle \mathbf{r}_1, \mathbf{r}_2, \ldots, \mathbf{r}_N | \mathcal{P} | nJM\rangle \equiv \langle -\mathbf{r}_1, -\mathbf{r}_2, \ldots, -\mathbf{r}_N | nJM\rangle \tag{9.8}$$

in which each spatial coordinate vector $\mathbf{r}_i(x_i, y_i, z_i)$ is represented by its Cartesian components. The parity operation changes the handedness of the coordinate system, but no continuous mathematical procedure can change the handedness of a coordinate system. Mechanical motion cannot accomplish sign inversion for all coordinates. Perhaps concerned that the term *discontinuous operator* is too harsh for inclusion in a legitimate physical theory, most call $\mathcal{P}$ a *discrete operator*. A discrete operator has the unusual feature of being both unity,

$$\mathcal{P}^{-1} = \mathcal{P}^\dagger \tag{9.9}$$

$$\mathcal{P}^{\dagger}\mathcal{P} = 1 \tag{9.10}$$

and Hermitian with an eigenvalue of ±1 [8–11],

$$\langle \mathbf{r}_1, \mathbf{r}_2, \ldots, \mathbf{r}_{\mathcal{N}}|\mathcal{P}|nJM\rangle = p\, \langle \mathbf{r}_1, \mathbf{r}_2, \ldots, \mathbf{r}_{\mathcal{N}}|nJM\rangle \tag{9.11a}$$

$$= \pm\langle \mathbf{r}_1, \mathbf{r}_2, \ldots, \mathbf{r}_{\mathcal{N}}|nJM\rangle. \tag{9.11b}$$

The determinant of the $\mathcal{P}$ matrix is -1 which classifies $\mathcal{P}$ as an improper rotation, that is, a rotation that mechanical rotation and coordinate rotation cannot accomplish. We will see below that coordinate rotation can invert the signs of two of the three components of $\mathbf{r}(x, y, z)$. In addition to being unitary and Hermitian, the parity operator is its own inverse,

$$\mathcal{P}^{-1} = \mathcal{P}. \tag{9.12}$$

The 3×3 matrix representation of $\mathcal{P}$ in equation (9.2) is easily seen to be Hermitian, unitary, and its own inverse.

There are one-dimensional parity operators, for example

$$\sigma_v(yz)\begin{bmatrix} x \\ y \\ z \end{bmatrix} = \begin{bmatrix} -1 & 0 & 0 \\ 0 & 1 & 0 \\ 0 & 0 & 1 \end{bmatrix}\begin{bmatrix} x \\ y \\ z \end{bmatrix} = \begin{bmatrix} -x \\ y \\ z \end{bmatrix}, \tag{9.13}$$

which is said to be 'reflection of the x-axis through the yz plane'. obfuscation which hides the simple mathematical facts that $\sigma_v(yz)$ inverts the sign on x but not on y and z. The determinant of the $\sigma_v(yz)$ matrix is clearly -1 marking it as a parity operator or improper rotation. There are also similar two-dimensional symmetry operations, for example

$$C_2(x)\begin{bmatrix} x \\ y \\ z \end{bmatrix} = \begin{bmatrix} 1 & 0 & 0 \\ 0 & -1 & 0 \\ 0 & 0 & -1 \end{bmatrix}\begin{bmatrix} x \\ y \\ z \end{bmatrix} = \begin{bmatrix} x \\ -y \\ -z \end{bmatrix}, \tag{9.14}$$

but here the determinant of the $C_2(x)$ matrix is $+1$, a proper rotation and not a parity operation. The product $\sigma_v(yz)$ and $C_2(x)$ matrices yield the $\mathcal{P}$ matrix of equation (9.2), and the operator $\mathcal{P}$ can be written as the product of operators,

$$\mathcal{P} = \sigma_v(yz)C_2(x). \tag{9.15}$$

As shown by, for example [12], parity is multiplicative. Angular momentum operators and eigenvalues are added, parity operators and eigenvalues are multiplied. The product $\sigma_v(yz)C_2(x)$ is appropriately applied to an eigenfunction depending on unprimed coordinates while $\sigma_v(y'z')C_2(x')$ is applied to an eigenfunction depending on rotated coordinates.

One can verify by direct substitution that the discrete transformations of the Euler angles

$$\alpha \to \pi + \alpha \tag{9.16a}$$

$$\beta \to \pi - \beta \tag{9.16b}$$

$$\gamma \to -\gamma \tag{9.16c}$$

change the signs of columns 2 and 3 in the 3 × 3 matrices in equations (H.6). With these discrete transformations of the Euler angles the parity operator becomes

$$\mathcal{P} = \sigma_v(yz)C_2(\pi + \alpha, \pi - \beta, -\gamma). \tag{9.17}$$

An exact formula for diatomic parity eigenvalues will be obtained below by applying the product $\sigma_v(yz)C_2(x)$ to the Wigner–Witmer eigenfunction.

9.4 Parity and angular momentum

Each wave in the spherical wave expansion of the electromagnetic wave carries specific values of parity and angular momentum [9, 13, 14]. Calculation of the transition moments of each multipole in the multipole expansion of the electromagnetic field yields selection rules for angular momentum and parity. For electromagnetic transitions, each angular momentum selection rule is paralleled by a parity selection rule. There is another connection between parity and angular momentum. The parity eigenvalue always separates into two parts, a constant part and an angular momentum dependent part. Before applying the parity operator to the Wigner–Witmer eigenfunction, let us get a better understanding of why the parity eigenvalue splits into two parts by first applying $\mathcal{P}$ to the very general equation

$$\langle \mathbf{r}_1, \mathbf{r}_2, \ldots, \mathbf{r}_N | \mathcal{R}(\alpha, \beta, \gamma) | nJM \rangle = \sum_{M'=-J}^{J} \langle \mathbf{r}'_1, \mathbf{r}'_2, \ldots, \mathbf{r}'_N | nJM' \rangle D^{J*}_{MM'}(\alpha, \beta, \gamma). \tag{9.18}$$

This equation essentially defines the Wigner D-function for coordinate (passive) rotations (see, e.g. equations (4.1.1)–(4.1.3) of [15]). The above equation describes the influence of coordinate rotation, not mechanical rotation, on an eigenfunction and this explains the appearance of the complex conjugate of $D^{J}_{MM'}(\alpha, \beta, \gamma)$.

When one forms the product of the parity operator $\mathcal{P}$ and rotation operator $\mathcal{R}(\alpha, \beta, \gamma)$, one has built an operator whose effect will depend on both parity and angular momentum. The angular momentum dependence comes because the components of angular momentum $\mathbf{J}$ are the generators of the rotations α, β, and γ (see, e.g. [11, 12, 16–20]).

Application of $\mathcal{P}$ to the left side of the general equation (9.18) and application $\sigma_v(y'z')C_2(\pi + \alpha, \pi - \beta, -\gamma)$ to the right yields

$$\begin{aligned}
\langle \mathbf{r}_1, \mathbf{r}_2, \ldots, \mathbf{r}_{\mathcal{N}} | \mathcal{P} | nJM \rangle
&= \sum_{M'=-J}^{J} \langle \mathbf{r}'_1, \mathbf{r}'_2, \ldots, \mathbf{r}'_{\mathcal{N}} | \sigma_v(y', z') | JM' \rangle D^{J^*}_{MM'}(\pi + \alpha, \pi - \beta, -\gamma) \\
&= p_\sigma (-)^{J+2M} \sum_{M'=-J}^{J} \langle -x'_1, -y'_1, -z'_1, -x'_2, -y'_2, -z'_2, \ldots, -x'_{\mathcal{N}}, \\
&\quad -y'_{\mathcal{N}}, -z'_{\mathcal{N}} | nJM \rangle D^{J^*}_{M,-M'}(\alpha, \beta, \gamma) \\
&= p \sum_{M'=-J}^{J} \langle \mathbf{r}'_1, \mathbf{r}'_2, \ldots, \mathbf{r}'_{\mathcal{N}} | JM' \rangle D^{J^*}_{M,-M'}(\alpha, \beta, \gamma)
\end{aligned} \tag{9.19}$$

or

$$p = p_\sigma (-)^{J+2M}. \tag{9.20}$$

This equation gives the general result that the parity eigenvalue splits into two parts, the angular momentum dependent part $(-)^{J+2M}$ and a constant part called the intrinsic parity, but there is a flaw in the angular momentum dependent part. A general result from angular momentum theory is that the $2J + 1$ states $|J, -J\rangle$, $|J, -J + 1\rangle$, …, $|J, J - 1\rangle$, $|JJ\rangle$ all have the same parity. Parity does not depend on the magnetic quantum number M. For integer angular momenta, the factor $(-)^{2M}$ in equation (9.20) simply disappears, but for half-integer momenta the M dependence introduces a minus sign,

$$p = -p_\sigma (-)^J \quad J \text{ half-integer}. \tag{9.21}$$

If one agrees to the convention to always subtract 1/2 from half-integer values of J, the above becomes

$$p = -p_\sigma (-)^{J-1/2} (-)^{1/2} \tag{9.22a}$$

$$= -ip_\sigma (-)^{J-1/2} \quad J \text{ half-integer}, \tag{9.22b}$$

and one sees that the intrinsic parity p_σ is purely imaginary when J is half-integer. If one further extends the convention to ignore the leading i on the right side of the above equation, one obtains

$$p = p_\sigma (-)^J \quad J \text{ integer} \tag{9.23a}$$

$$= -p_\sigma (-)^{J-1/2} \quad J \text{ half-integer} \tag{9.23b}$$

in which all terms can be regarded as real.

9.5 Diatomic parity

Operation of $\sigma_v(y'z')C_2(\pi + \alpha, \pi - \beta, -\gamma)$ on the right side of equation (3.16) givesright side of equation

$$\begin{aligned}
&\sum_{\Omega=-J}^{J} \langle \rho, \zeta, \mathbf{r}'_2, \ldots, \mathbf{r}'_N, r|\sigma_v(y'z')|nv\rangle D^{J^*}_{M\Omega}(\pi+\phi, \pi-\theta, -\chi) \\
&= \sum_{\Omega=-J}^{J} \langle \rho, -\zeta, -x'_2, -y'_2, -z'_2, \ldots, -x'_N, -y'_N, -z'_N, r|nv\rangle D^{J^*}_{M,-\Omega}(\phi, \theta, \chi) \\
&= p_\Sigma(-)^{J+2\Omega} \sum_{\Omega=-J}^{J} \langle \rho, \zeta, \mathbf{r}'_2, \ldots, \mathbf{r}'_N, r|nv\rangle D^{J^*}_{M,-\Omega}(\phi, \theta, \chi) \\
&= p \sum_{\Omega=-J}^{J} \langle \rho, \zeta, \mathbf{r}'_2, \ldots, \mathbf{r}'_N, r|nv\rangle D^{J^*}_{M\Omega}(\phi, \theta, \chi).
\end{aligned} \tag{9.24}$$

In the last line, the minus sign on Ω in $D^{J^*}_{M\Omega}(\phi, \theta, \chi)$ was dropped because changing this sign merely reverses the order in which the sum over Ω is formed. Comparison of the last two lines gives

$$p = p_\Sigma(-)^{J+2\Omega} \tag{9.25}$$

in which p_Σ is the intrinsic parity of the diatomic molecule. The conventions that lead from equation (9.18) to equation (9.23) give

$$p = p_\Sigma(-)^J \quad J \text{ integer} \tag{9.26a}$$

$$= -\, p_\Sigma(-)^{J-1/2} \quad J \text{ half-integer} \tag{9.26b}$$

9.6 Λ doublets

The separation of ϕ, θ, and χ becomes apparent when $D^{J^*}_{M\Omega}(\phi, \theta, \chi)$ is broken into it parts,

$$D^{J^*}_{M\Omega}(\phi, \theta, \chi) = e^{iM\phi} d^{J}_{M\Omega}(\beta) e^{i\Omega\chi} \tag{9.27}$$

We saw above that the diatomic parity operator changes the sign of χ. From equations (9.6), one sees that the sum over Ω from $-J$ to J in the Wigner–Witmer eigenfunction (3.16) guarantees that each diatomic state of a given parity has a parity twin.

References

[1] Hornkohl J O and Parigger C G 2017 *Int. J. Mol. Theor. Phys.* **1** 00103
[2] Hornkohl J O and Parigger C G 2017 *Int. J. Mol. Theor. Phys.* **1** 00102
[3] Hougen J T 1962 *J. Chem. Phys.* **36** 519
[4] Larsson M 1981 *Phys. Scr.* **23** 835
[5] Zare R N 1988 *Angular Momentum* (New York: Wiley)
[6] Lefebvre-Brion H and Field R W 2004 *The Spectra and Dynamics of Diatomic Molecules* (New York: Elsevier/Academic)

[7] Brown J M and Carrington A 2003 *Rotational Spectroscopy of Diatomic Molecules* (Cambridge: Cambridge University Press)
[8] Kemble E C 2005 *The Fundamental Principles of Quantum Mechanics* (Mineola, NY: Dover)
[9] Messiah A 1964 *Quantum Mechanics* (Amsterdam: North Holland)
[10] Cohen-Tannoudji C, Diu B and Laloë F 1977 *Quantum Mechanics* (New York: Wiley)
[11] Zettili N 2009 *Quantum Mechanics* 2nd edn (Chichester: Wiley)
[12] Gottfried K and Yan T-M 1998 *Quantum Mechanics: Fundamentals* 2nd edn (New York: Springer)
[13] Rose M E 1995 *Elementary Theory of Angular Momentum* (Mineola, NY: Dover)
[14] Edmonds A R 1974 *Angular Momentum in Quantum Mechanics* 2nd edn (Princeton, NJ: Princeton University Press)
[15] Varshalovich D A, Moskalev A N and Khersonskii V K 1988 *Quantum Theory of Angular Momentum* (Singapore: World Scientific)
[16] Landau L D and Lifshitz E M 1977 *Quantum Mechanics* 3rd edn (Amsterdam: Butterworth-Heinemann)
[17] Tung W-K 1985 *Group Theory in Physics* (Singapore: World Scientific)
[18] Merzbacher E 1998 *Quantum Mechanics* 3rd edn (New York: Wiley)
[19] Chaichain M and Hagedorn R 1998 *Symmetries in Quantum Mechanics* (New York: Taylor & Francis)
[20] Weinberg S 2015 *Lectures on Quantum Mechanics* 2nd edn (Cambridge: Cambridge University Press)

Christian G Parigger and James O Hornkohl

Chapter 10

The Condon and Shortley line strength

The treatment of the diatomic molecule leads to line strengths that are utilized for the computation of molecular spectra. The *strength* of a line [1], or the line strength, is introduced in this chapter.

The geometric configuration of the charged electrons and nuclei in the molecule produces electric and magnetic moments. In a molecule governed by classical mechanics and electrodynamics, molecular motion would produce multipole radiation (and the rapid collapse of the molecule). In quantum mechanics a multipole of charge radiates only during quantum jumps to a lower energy level and absorbs radiation only during a quantum jump to a higher level. The total Hamiltonian for the system of molecule plus electromagnetic field is the sum of a quantized molecular Hamiltonian and a quantized electromagnetic field.

Emission and absorption of light is produced by creation and annihilation operators acting on a quantized electromagnetic field. The quantum mechanical model being used here is that of a molecule residing in the always present quantized radiation field. This model of molecule plus field is conservative. Energy and momentum gained by the molecule are lost by the field and vice versa. The molecule and radiation field interact only during quantum jumps, and each can be treated as an independent conservative system before and after quantum jumps.

A multipole expansion can be made of the free electromagnetic field. Transition matrix elements (transition moments) of the individual terms $T_k^{(q)}$ in the multipole expansion of the electromagnetic field control the probability of radiative transitions, and the source of selections rules.

Condon and Shortley write in the preface to their book [1] 'We have defined in §7[4] a quantity, called *strength* of a line, which we find to give a more convenient theoretical specification of the radiation intensity than either of the Einstein transition probabilities. We hope that this new usage will find favour among

doi:10.1088/978-0-7503-1989-8ch10

spectroscopists'. In the following, the *strength* definition is stated as provided by Condon and Shortley [1].

The line strength [1] is introduced by considering first the actual emitted intensity of the component from state a to state b,

$$I(a, b) = N(a)h\nu\mathbf{A}(a, b), \tag{10.1}$$

where $N(a)$ is the number of atoms in state a and $\mathbf{A}(a, b)$ is the spontaneous emission probability for that transition. This equation holds only when the radiation density present is so small that the induced emission is negligible.

The total *intensity* of a line is the sum of the intensities of its components, hence in natural excitation (sufficiently isotropic excitation) for the line from level A to B,

$$I(A, B) = N(a)h\nu\frac{64\pi^4\sigma^3}{3h}\sum_{a,b}|(a|\mathbf{P}|b)|^2, \quad \sigma = \frac{\nu}{c} = \tilde{\nu}, \tag{10.2}$$

in the case of dipole radiation. The sum of the squared matrix components occurring here we shall define as the *strength* of the line, $\mathbf{S}(A, B)$, i.e. sum over a and b. The intensity of a particular line is therefore proportional to the number of atoms in any one of the initial states, to σ^4 and to $\mathbf{S}(A, B)$.

The intensity is written as proportional to the total number of atoms in the initial level, $N(A)$,

$$I(A, B) = N(A)h\nu\mathbf{A}(A, B). \tag{10.3}$$

As $N(A) = (2j_A + 1)N(a)$, this gives

$$\mathbf{A}(A, B) = \frac{1}{2j_A + 1}\frac{64\pi^4\sigma^3}{3h}\mathbf{S}(A, B). \tag{10.4}$$

Here, $\mathbf{A}$ is the spontaneous-transition probability of Einstein. Of course, in SI-units and conventional writing of wave-numbers, $\tilde{\nu}$, equation (10.4) reads

$$\mathbf{A}(A, B) = \frac{1}{2j_A + 1}\frac{16\pi^3\tilde{\nu}^3}{3\varepsilon_0 h}\mathbf{S}(A, B). \tag{10.5}$$

Reference

[1] Condon E U and Shortley G 1953 *The Theory of Atomic Spectra* (Cambridge: Cambridge University Press)

IOP Publishing

Quantum Mechanics of the Diatomic Molecule with Applications

Christian G Parigger and James O Hornkohl

Chapter 11

Hönl–London line-strength factors in Hund's cases (a) and (b)

11.1 Case (a) basis functions

Hund's case (a) basis function is obtained when one drops the summation in the Wigner–Witmer eigenfunction (3.16), re-normalizes the result, and inserts the spin ket $|S\Sigma\rangle$,

$$
\begin{aligned}
|a\rangle &= \langle \mathbf{r}'_1, \mathbf{r}'_2, \ldots, \mathbf{r}'_{N-1}, r, \rho, \zeta, \phi, \theta, \chi \, | nvJM\Omega\Lambda S\Sigma \rangle \\
&= \sqrt{\frac{2J+1}{8\pi^2}} \, \langle \mathbf{r}'_1, \mathbf{r}'_2, \ldots, \mathbf{r}'_{N-1}, r, \rho, \zeta \, | nv \rangle \, | S\Sigma \rangle \, D^{J^*}_{M\Omega}(\phi, \theta, \chi).
\end{aligned} \tag{11.1}
$$

If nuclear spin is ignored and one writes the total angular momentum $\mathbf{J}$ in terms of the total orbital angular momentum $\mathbf{N}$,

$$\mathbf{N} = \mathbf{L} + \mathbf{R} \tag{11.2}$$

where $\mathbf{L}$ is the total orbital angular momentum of the electrons, $\mathbf{R}$ is the orbital angular momentum of the nuclei, and $\mathbf{S}$ is the total spin of the electrons,

$$\mathbf{J} = \mathbf{N} + \mathbf{S} \tag{11.3}$$

and then makes the inverse Clebsch–Gordan series expansion of $D^{J^*}_{M\Omega}(\phi, \theta, \chi)$,

$$
\begin{aligned}
D^{J^*}_{M\Omega}(\phi, \theta, \chi) = &\sum_{M_N=-N}^{N} \; \sum_{M_S=-S}^{S} \; \sum_{\Sigma=-S}^{S} \; \sum_{\Lambda=-N}^{N} \\
&\times \langle NM_N; SM_S | JM \rangle \, \langle N\Lambda; S\Sigma | J\Omega \rangle \\
&\times D^{N^*}_{M_N\Lambda}(\phi, \theta, \chi) \, D^{S^*}_{M_S\Sigma}(\phi, \theta, \chi)
\end{aligned} \tag{11.4}
$$

one sees that the same set of Euler angles appear in $D^{N*}_{M_N\Lambda}(\phi, \theta, \chi)$ and $D^{S*}_{M_S\Sigma}(\phi, \theta, \chi)$ indicating that the z' component of **S** (quantum number Σ) and z' component of **L** (quantum number Λ) are perfectly coupled, and the quantum number Ω is the sum of Λ and Σ,

$$\Omega = \Lambda + \Sigma. \tag{11.5}$$

11.2 Case (b) basis functions

In Hund's case (b), the electronic spin states are independent of the electronic orbital angular momentum states. The $|JM\rangle$ states cannot be built using the inverse Clebsch–Gordan series because the spin states do not rotate in sync with the orbital rotations. The case (b) eigenfunction can be built by coupling $|NM_N\rangle$ states with $|SM_S\rangle$ states with the Clebsch–Gordan coefficient $\langle NM_N;\, SM_S\,|JM\rangle$. The $|NM_N\rangle$ states are obtained by replacing the quantum numbers J, M, and Ω in the Wigner–Witmer eigenfunction with N, M_N, and Λ, dropping the summation, and renormalizing the result. When this spin free result is coupled to the electronic spin states one obtains $|SM_S\rangle$

$$\begin{aligned} |b\rangle &= \langle \mathbf{r}'_1, \mathbf{r}'_2, \ldots, \mathbf{r}'_{N-1}, r, \rho, \zeta, \phi, \theta, \chi\ |nvJMN\Lambda S\rangle \\ &= \sqrt{\frac{2N+1}{8\pi^2}} \sum_{M_N=-N}^{N} \sum_{M_S=-S}^{S} \\ &\quad \times \langle \mathbf{r}'_1, \mathbf{r}'_2, \ldots, \mathbf{r}'_{N-1}, r, \rho, \zeta\ |n\rangle\ \langle NM_N;\, SM_S|JM\rangle \\ &\quad \times |SM_S\rangle\ D^{N*}_{M_N\Lambda}(\phi, \theta, \chi). \end{aligned} \tag{11.6}$$

11.3 Mathematical properties of case (a) and case (b) basis functions

The case (a) and case (b) basis functions are complete, orthogonal eigenfunctions. However, they do not possess the mathematical properties of quantum mechanical angular momentum states. For example, the familiar behavior of angular momentum states under operation by the raising and lowering operators,

$$J_{\pm}|JM\rangle = C_{\pm}(JM)\,|J,\, M \pm 1\rangle \tag{11.7}$$

$$J_{\pm}|J\Omega\rangle = C_{\pm}(J\Omega)\,|J,\, \Omega \pm 1\rangle \tag{11.8}$$

where

$$C_{\pm}(JM)\ \sqrt{J(J+1) - M(M \pm 1)} \tag{11.9}$$

is not duplicated when the raising and lowering operators are applied to rotation matrix elements

$$J_{\pm}D^{J}_{M\Omega}(\phi, \theta, \chi) = -\sqrt{J(J+1) - M(M \mp 1)}\ D^{J}_{M\mp 1,\Omega}(\phi, \theta, \chi) \tag{11.10}$$

$$J_{\pm}D^{J^*}_{M\Omega}(\phi, \theta, \chi) = \sqrt{J(J+1) - M(M \pm 1)}\; D^{J^*}_{M\pm 1, \Omega}(\phi, \theta, \chi) \tag{11.11}$$

$$J'_{\pm}D^{J}_{M\Omega}(\phi, \theta, \chi) = \sqrt{J(J+1) - \Omega(\Omega \pm 1)}\; D^{J}_{M, \Omega\pm 1}(\phi, \theta, \chi) \tag{11.12}$$

$$J'_{\pm}D^{J^*}_{M\Omega}(\phi, \theta, \chi) = -\sqrt{J(J+1) - \Omega(\Omega \mp 1)}\; D^{J^*}_{M, \Omega\mp 1}(\phi, \theta, \chi) \tag{11.13}$$

One sees from the above that raising operators sometimes lower the magnetic quantum number and sometimes introduce an unexpected sign change. Whereas in an angular momentum state, the total and one component of the total are constants of the motion, in a rotation matrix element basis function the total and *two* of it components are constants.

$$J_z D^{J^*}_{M\Omega}(\phi, \theta, \chi) = J_z e^{iM\phi} d^{J}_{M\Omega}(\theta)\, e^{i\Omega\chi} \tag{11.14}$$

$$= -i\frac{\partial}{\partial\phi}\, D^{J^*}_{M\Omega}(\phi, \theta, \chi) \tag{11.15}$$

$$= M\; D^{J^*}_{M\Omega}(\phi, \theta, \chi) \tag{11.16}$$

$$J_{z'} D^{J^*}_{M\Omega}(\phi, \theta, \chi) = J_{z'} e^{iM\phi} d^{J}_{M\Omega}(\theta)\, e^{i\Omega\chi} \tag{11.17}$$

$$= -i\frac{\partial}{\partial\chi}\, D^{J^*}_{M\Omega}(\phi, \theta, \chi) \tag{11.18}$$

$$= \Omega\; D^{J^*}_{M\Omega}(\phi, \theta, \chi) \tag{11.19}$$

A self-evident but mathematically rigorous explanation of why $D^{J^*}_{M\Omega}(\phi, \theta, \chi)$ does not represent a state of angular momentum is that it carries two magnetic quantum numbers. As the general equation equation (3.8) and Wigner–Witmer equation (3.16) show, one must sum the rotation matrix elements over one of the extra magnetic quantum numbers to obtain a true angular momentum eigenstate.

11.4 Diatomic parity operator

The diatomic parity operator is the product to two parity operators because Euler rotations are proper but the parity operator (inversion of the sign of spatial coordinates) is an improper rotation (i.e. one that changes the handedness of the coordinate system and therefore cannot be the result of physical rotation.) Three different transformations of the Euler angles can change the sign of two of the three components of a molecule-fixed spatial coordinate vector. The transformations $\phi \to \pi + \phi, \theta \to \pi - \theta$ and $\chi \to -\chi$ change the signs of the y' and x' components of all molecule-fixed spatial vectors. Thus, if one defines the operator P_Σ to invert the signs of the x' components, the parity diatomic parity operator can be written as

$$P = P_\Sigma P(\phi, \theta, \chi). \tag{11.20}$$

Scaler invariants (distances) such as r and ρ, and spin which cannot be expressed in terms of spatial coordinates are immune to the parity operator. Application of the parity operator to the case (a) basis yields

$$
\begin{aligned}
&P_\Sigma P(\phi, \theta, \chi)\sqrt{\frac{2J+1}{8\pi^2}}\langle \mathbf{r}_1', \mathbf{r}_2', \ldots, \mathbf{r}_{N-1}', r, \rho, \zeta\,|nv\rangle\, D^{J^*}_{M\Omega}(\phi, \theta, \chi) \\
&= \sqrt{\frac{2J+1}{8\pi^2}}\langle -\mathbf{r}_1', -\mathbf{r}_2', \ldots, -\mathbf{r}_{N-1}', r, \rho, \zeta\,|n\rangle\, D^{J^*}_{M,-\Omega}(\phi, \theta, \chi) \qquad (11.21)\\
&= p_\Sigma(-)^{J+2M}\sqrt{\frac{2J+1}{8\pi^2}}\langle \mathbf{r}_1', \mathbf{r}_2', \ldots, \mathbf{r}_{N-1}', r, \rho, \zeta\,|nv\rangle\, D^{J^*}_{M,-\Omega}(\phi, \theta, \chi).
\end{aligned}
$$

When applied to a parity state, the parity operator can at most change the sign of the eigenfunction, but as seen above the parity operator converts $D^{J^*}_{M\Omega}(\phi, \theta, \chi)$ into the orthogonal function $D^{J^*}_{M,-\Omega}(\phi, \theta, \chi)$. A rotation matrix element lacks the mathematical properties of a parity eigenfunction. Thus, the case (a) and case (b) basis functions are not parity eigenfunctions.

Our molecular model here includes only the electromagnetic field. In particular, the weak force is excluded. Parity is a member of the collection of commuting observables. Also, the multipole expansion of the electromagnetic field is an eigenfunction of the total angular momentum, one of its components, and parity [1]. In summary, angular momentum and parity are in perfect accord. It comes as no surprise that if $D^{J^*}_{M\Omega}(\phi, \theta, \chi)$ is not a parity state then it is also not an angular momentum state and vice versa.

Matrix elements of the parity operator is Hund's cases (a) and (b) are given by

$$p^{(a)}_{ij} = p_\Sigma\,(-)^J\,\delta(J_iJ_j)\,\delta(\Omega_i, -\Omega_j)\,\delta(\Lambda_i, -\Lambda_j)\,\delta(n_i\,n_j) \qquad (11.22a)$$

$$p^{(b)}_{ij} = p_\Sigma\,(-)^{N_i}\,\delta(N_i\,N_j)\,\delta(\Lambda_i, -\Lambda_j)\,\delta(n_i\,n_j) \qquad (11.22b)$$

11.5 Hönl–London line-strength factors

The Hönl–London line-strength factor, sometimes called the rotational line-strength factor, is the angular momentum part of the Condon and Shortley line strength for a diatomic molecule. The probability of an exchange of energy between a molecule and the electromagnetic field (i.e. the probability of a radiative transition) is controlled by the matrix elements of the operators for the interactions between the molecule and radiation field. Typically, the matrix elements of one tensor operator $\mathbf{T}^{(q)}$ are much larger than the matrix elements for all other operators, and the line strength is defined as the absolute square of the matrix element $\langle f|\mathbf{V}|i\rangle$ summed over all transitions leading to the same observation (e.g. same spectral line.).

It was noted above that equations (3.8) and (3.16) merely answer the question of what happens to the eigenfunction when coordinate changes the direction of the

z-axis. One can ask the same question of operators. An operator that transforms in the same way as angular momentum eigenfunctions transform is called a spherical tensor operator. For example, the spherical tensor components of the vector operator $\mathbf{V}(xyz)$ are defined as

$$T_1^{(1)} = -\sqrt{\frac{1}{2}}\left(V_x + iV_y\right) \tag{11.23}$$

$$T_0^{(1)} = V_z \tag{11.24}$$

$$T_{-1}^{(1)} = \sqrt{\frac{1}{2}}\left(V_x - iV_y\right) \tag{11.25}$$

and its components in the laboratory coordinate system are related to those in the rotated coordinate system by

$$T_k^{(1)}(xyz) = \sum_{\kappa=-1}^{1} T_\kappa^{(1)}(x'y'x')\, D_{k\kappa}^{1^*}(\phi, \theta, \chi). \tag{11.26}$$

One sees that this is just equation (3.16) written for the vector operator $\mathbf{V}(xyz)$ instead of for the diatomic eigenfunction.

11.6 Triple integral of three rotation matrix elements

$$\begin{aligned} &\int_0^{2\pi}\int_0^{\pi}\int_0^{2\pi} D_{m_3\omega_3}^{j_3^*}(\phi, \theta, \chi)\, D_{m_2\omega_2}^{j_2}(\phi, \theta, \chi)\, D_{m_1\omega_1}^{j_1}(\phi, \theta, \chi)\sin(\theta)\, d\phi\, d\theta\, d\chi \\ &= \frac{8\pi^2}{2j_3 + 1}\langle j_1\ m_1;\, j_2\ m_2 | j_3\ m_3\rangle\, \langle j_1\ \omega_1;\, j_2\ \omega_2 | j_3\ \omega_3\rangle \end{aligned} \tag{11.27}$$

The Clebsch–Gordan coefficients are real. Therefore, the above integral is real. From the formula relating the rotation matrix element to its complex conjugate,

$$D_{j\omega}^{j}(\alpha, \beta, \gamma) = (-)^{\omega - m} D_{-m,-\Omega}^{j^*}(\alpha, \beta, \gamma), \tag{11.28}$$

it is seen that complex conjugation of one or more of the rotation matrix elements might change the sign of the integral.

11.7 Calculation of the Hönl–London line-strength factors for cases (a) and (b)

Given here are mathematically simple (and somewhat simple minded) calculations of Hönl–London line-strength factors for cases (a) and (b). The calculation is described as somewhat simple-minded because one knows that any quantum state must possess the mathematical properties of angular momentum and if the system is controlled by electromagnetic forces it must also possess parity. The basis function equations(11.1) and (11.6) possess neither.

Placing the case (a) basis functions into the definition of the line strength (10) gives

$$S(n'J', nJ) = \sum_{M=-J}^{J} \sum_{M'=-J'}^{J'} \sum_{k=-q}^{q} \sum_{\kappa=-q}^{q}$$

$$\left| \frac{1}{8\pi^2} \sqrt{(2J+1)(2J'+1)} \langle nv | T_{k\kappa}^{(q)}(\mathbf{r}_1', \mathbf{r}_2', \ldots, \mathbf{r}_{N-1}', r, \rho, \zeta) | n'v' \rangle \right. \tag{11.29}$$

$$\left. \int_0^{2\pi} \int_0^{2\pi} \int_0^{2\pi} D_{M\Omega}^{J}(\phi, \theta, \chi)\, D_{k\kappa}^{q^*}(\phi, \theta, \chi)\, D_{M'\Omega'}^{J'^*}(\phi, \theta, \chi) \sin(\theta)\, d\phi\, d\theta\, d\chi \right|^2 .$$

Whereas in the standard integral (11.27), the complex conjugate of only one Wigner D-function appears, in the above two D-functions appear as their complex conjugates. As noted above, this can only change the sign of the integral but because the integral will be squared we can ignore the possible sign change. The integral signs become vitally important below where sums of the integrals are squared. Evaluation of the triple integral above gives

$$\begin{aligned}
S(n'v'J', nvJ) &= \frac{2J'+1}{2J+1} \sum_{M=-J}^{J} \sum_{M'=-J'}^{J'} \sum_{k=-q}^{q} \sum_{\kappa=-q}^{q} \\
&\times \left| \left\langle nv \,| T_{k\kappa}^{(q)} \,|(\mathbf{r}_1'\mathbf{r}_2' \ldots \mathbf{r}_{N-1}' \, r \, \rho \, \zeta) | n'v' \right\rangle \langle J' \, M'; \, q \, k \,| J \, M \rangle \langle J' \Omega'; \, q \, \kappa \,| J \, \Omega \rangle \right|^2 \\
&= \frac{2J'+1}{2J+1} \sum_{\kappa=-q}^{q} \left| \left\langle nv | T_{k\kappa}^{(q)}(\mathbf{r}_1', \mathbf{r}_2', \ldots, \mathbf{r}_{N-1}', r, \rho, \zeta) | n'v' \right\rangle \right|^2 \\
&\times \langle J' \Omega'; \, q \, \kappa \,| J \, \Omega \rangle^2 \sum_{M=-J}^{J} 1 \\
&= (2J'+1) \left| \left\langle nv \,| T_{k,\Omega'-\Omega}^{(q)} | n'v' \rangle (\mathbf{r}_1'\mathbf{r}_2' \ldots \mathbf{r}_{N-1}' \, r \, \rho \, \zeta) | n' \right\rangle \right|^2 \\
&\times \langle J' \Omega' \, q, \, \Omega - \Omega' | J \, \Omega \rangle^2 \\
&= S(n'v', nv)\, S(J', J)
\end{aligned} \tag{11.30}$$

where

$$S(n'v', nv) = \left| \sum_{k=-q}^{q} \langle n | T_{k,\Omega'-\Omega}^{(q)}(\mathbf{r}_1', \mathbf{r}_2', \ldots, \mathbf{r}_{N-1}', r, \rho, \zeta) \, | n'v' \rangle \right|^2 \tag{11.31}$$

$$S(J', J) = (2J'+1) \langle J' \Omega'; \, q, \, \Omega - \Omega' \, | J \, \Omega \rangle^2 \tag{11.32}$$

In the above, we used

$$\sum_{m_3=-j_3}^{j_3}\sum_{m_1=-j_1}^{j_1}\sum_{m_2=-j_2}^{j_2}\langle j_1\, m_1; j_2\, m_2\, | j_3\, m_3\rangle\langle j_1\, m_1; j_2\, m_2\, | j_3'\, m_3'\rangle = \sum_{m_3=-j_3}^{j_3}\delta_{j_3 j_3'}\delta_{m_3 m_3'} = 2j_3 + 1. \tag{11.33}$$

in

$$\sum_{M=-J}^{J}\sum_{M'=-J'}^{J'}\sum_{k=-q}^{q}\langle J'\, M'; q\, k\, | J\, M\rangle\langle J'\, M'; q\, k\, | J\, M\rangle = \sum_{M=-J}^{J}\delta_{JJ}\delta_{MM} = 2J + 1. \tag{11.34}$$

We could have used the symmetry of the Clebsch–Gordan coefficients in which $j_2\ m_2$ and $j_3\ m_3$ are exchanged to write

$$\begin{aligned}\sum_{m_3=-j_3}^{j_3}\sum_{m_1=-j_1}^{j_1}\sum_{m_2=-j_2}^{j_2}&\langle j_1\, m_1; j_2\, m_2\, | j_3\, m_3\rangle\langle j_1\, m_1; j_2\, m_2\, | j_3\, m_3\rangle\\ &=\left[(-)^{j_1-m_1}\sqrt{\frac{2j_3+1}{2j_2+1}}\right]^2\sum_{m_2=-j_2}^{j_2}\sum_{m_1=-j_1}^{j_1}\sum_{m_3=-j_3}^{j_3}\langle j_1\, m_1; j_3, -m_3\, | j_2, -m_2\rangle^2\\ &=\frac{2j_3+1}{2j_2+1}\sum_{m_2=-j_2}^{j_2}\delta_{j_2 j_2}\delta_{m_2 m_2}\\ &=2j_3+1\end{aligned} \tag{11.35}$$

The line strength is symmetrical with respect to the upper and lower levels,

$$S\left(n'v_{n'}'J', nv_nJ\right) = S\left(nv_nJ, n'v_{n'}'J'\right) \tag{11.36}$$

and the Hönl–London line-strength factor should share this symmetry. From two symmetries of the Clebsch–Gordan coefficient, the first which swaps $\langle j_1 m_1|$ and $|j_3 m_3\rangle$,

$$\langle j_1\, m_1; j_2\, m_2\, | j_3\, m_3\rangle = \sqrt{\frac{2j_3+1}{2j_1+1}}\,(-)^{j_1-m_1+j_3-m_3}\langle j_3\, m_3; j_2\, m_2\, | j_1\, m_1\rangle, \tag{11.37}$$

and the second which changes the signs of all three magnetic quantum numbers,

$$\langle j_1\, m_1; j_2\, m_2\, | j_3\, m_3\rangle = (-)^{j_1-j_2-j_3}\langle j_1, -m_1; j_2, -m_2\, | j_3, -m_3\rangle, \tag{11.38}$$

and because the case (a) Hönl–London line-strength factor, equation(11.32), is expressed as the square of the Clebsch–Gordan coefficient, the Hönl–London line-strength factors for case (a) to case (a) transitions are symmetrical as required,

$$(2J' + 1)\,\langle J'\Omega';\, q,\, \Omega - \Omega'\,|J\,\Omega\rangle^2 = (2J + 1)\langle J\,\Omega;\, q,\, \Omega' - \Omega\,|J'\Omega'\rangle^2. \quad (11.39)$$

11.8 Hund's case (b) Hönl–London line-strength factors

The probability of the transition $nvJM \leftrightarrow n'v'J'M'$ is controlled by the matrix elements $\langle nvJM\,|T^{(q)_k}|n'v'J'M'\rangle$, sometimes called transition moments, where $T_k^{(q)}$ is the kth component of the tensor of degree q responsible for the transition. If the system in question depends upon the internal variables $\rho,\,\zeta,\,\chi,\,\mathbf{r}_2,\,\dots,\,\mathbf{r}_N,\,r,\,\theta,\,\phi$, then the operators responsible for the electromagnetic transitions (e.g. electric dipole, magnetic dipole, electric quadrupole, *etc.*) will depend upon the same coordinates as the eigenfunction Φ_{nvJM}, and when expressed as an irreducible tensor, will have the same behavior under coordinate rotation,

$$T_k^{(q)}(\rho,\,\zeta,\,\chi,\,\mathbf{r}_2,\,\dots,\,\mathbf{r}_N,\,r,\,\theta,\,\phi) = \sum_{\kappa=-q}^{q} T^{(q)}(\rho,\,\zeta,\,\mathbf{r}_2,\,\dots,\,\mathbf{r}_N,\,r)\,D_{k\kappa}^{q^*}(\phi,\,\theta,\,\chi) \quad (11.40)$$

as the eigenfunction Φ_{nvJM}. For Hund's case (b) basis functions, the transition moments become

$$\begin{aligned}
\langle b\,|T_k^{(q)}|b'\rangle &= \sum_{k=-q}^{q}\sum_{\kappa=-q}^{q}\left\langle nv\left|T_\kappa^{(q)}(\rho,\,\zeta,\,\mathbf{r}'_2,\,\mathbf{r}'_N,\,r)\right|n'v'\right\rangle \\
&\times \frac{\sqrt{(2N + 1)(2N' + 1)}}{8\pi^2}\sum_{M_N-N}^{N}\sum_{M_S=-S}^{S}\sum_{M'_N=-N'}^{N'}\sum_{M'_S=-S'}^{S'}\langle SM_S\,|S'M'_S\rangle \\
&\times \langle NM_N;\,SM_S\,|JM\rangle\,\langle N'M'_N;\,SM_S\,|J'M'\rangle \\
&\times \int_0^{2\pi}\int_0^{\pi}\int_0^{2\pi} D_{M_N\Lambda}^{N^*}(\phi,\,\theta,\,\chi)\,D_{k\kappa}^{q}(\phi,\,\theta,\,\chi) \\
&\times D_{M'_N\Lambda'}^{N'}(\phi,\,\theta,\,\chi)\sin(\theta)\,d\phi\,d\theta\,d\chi \\
&= \sqrt{2N + 1}\sum_{\kappa=-q}^{q}\left\langle nv\left|T_\kappa^{(q)}(\rho,\,\zeta,\,\mathbf{r}'_2,\,\mathbf{r}'_N,\,r)\right|n'v'\right\rangle\sum_{M_N=N}^{N}\sum_{M_S=-S}^{S}\sum_{M'_N=-N'}^{N'} \\
&\times \langle NM_N;\,SM_S\,|JM\rangle\,\langle N'M'_N;\,SM_S\,|J'M'\rangle \\
&\times \langle NM_N;\,q\,k\,|N'M'_N\rangle\,\langle N\Lambda;\,q\,\kappa\,|N'\Lambda'\rangle \\
&= \sqrt{2N + 1}\sum_{\kappa=-q}^{q}\sum_{M_N=N}^{N}\sum_{M_S=-S}^{S}\sum_{M'_N=-N'}^{N'}\left\langle nv\left|T_\kappa^{(q)}(\rho,\,\zeta,\,\mathbf{r}'_2,\,\mathbf{r}'_N,\,r)\right|n'v'\right\rangle \\
&\times \langle NM_N;\,SM_S\,|JM\rangle\,\langle N'M'_N;\,SM_S\,|J'M'\rangle \\
&\times \langle NM_N;\,q\,k\,|N'M'_N;\,\rangle\,\langle N\Lambda;\,q\,\kappa\,|N'\Lambda'\rangle \\
&= S(n'v',\,nv)\,S(J',\,J)
\end{aligned} \quad (11.41)$$

$$S(J', J) = (2J+1)(2J'+1)(2N+1)\langle N\Lambda\, q\,\kappa|N'\Lambda'\rangle^2 \begin{Bmatrix} q & J' & J \\ S & N & N' \end{Bmatrix}^2 \delta(S', S) \quad (11.42)$$

$$\begin{aligned} &\sum_{\kappa=-q}^{q} \sum_{M_N=N}^{N} \sum_{M_S=-S}^{S} \sum_{M'_N=-N'}^{N'} \langle NM_N SM_S\,|JM\rangle\, \langle N'M'_N SM_S\,|J'M'\rangle \\ &\times \langle NM_N\, q\, k\,|N'M'_N\rangle\, \langle N\Lambda\, q\,\kappa\,|N'\Lambda'\rangle \\ &= \sqrt{(2J'+1)(2J+1)} \begin{Bmatrix} q & J' & J \\ S & N & N' \end{Bmatrix} \end{aligned} \quad (11.43)$$

$$\begin{Bmatrix} a & b & c \\ d & e & f \end{Bmatrix} = \begin{Bmatrix} d & e & c \\ a & b & f \end{Bmatrix} \quad (11.44)$$

$$W(abcd;\, ef) = (-)^{b+e-c-f}\, W(afcd;\, eb) \quad (11.45)$$

$$\begin{Bmatrix} q & J' & J \\ S & N & N' \end{Bmatrix} = (-)^{J'+J-N-N'} \begin{Bmatrix} q & N' & J \\ S & N & J' \end{Bmatrix} \quad (11.46)$$

$$S(J', J) = (2J+1)(2J'+1)(2N+1)\langle N\Lambda\, q\,\kappa|N'\Lambda'\rangle^2 \begin{Bmatrix} q & N' & J \\ S & N & J' \end{Bmatrix}^2 \delta(S', S) \quad (11.47)$$

$$\begin{Bmatrix} j_1 = q & j_2 = N' & j_3 = J \\ j_4 = S & j_5 = N & j_6 = J' \end{Bmatrix} \quad (11.48)$$

$$\begin{Bmatrix} j_1 = q & j_2 = N' & J_{12} = J \\ j_3 = S & j = N & j_{23} = J' \end{Bmatrix} \quad (11.49)$$

$$\begin{Bmatrix} a = q & b = N' & e = J \\ d = S & c = N & f = J' \end{Bmatrix} \quad (11.50)$$

11.9 The electronic–vibrational strength

The Hönl–London line-strength factors are unitless. The electronic–vibrational strength $S(n'v', nv)$ carries the units of the Condon and Shortley line strength. The HITRAN line strength is not the Condon and Shortley line strength but can be expressed in terms of it.

The Born–Oppenheimer approximation separates the electronic–vibrational eigenfunction into the product of electronic and vibrational eigenfunctions,

$$\langle \mathbf{r}'_1, \mathbf{r}'_2, \ldots, \mathbf{r}'_{N-1}, r, \rho, \zeta\,|nv\rangle \approx \langle \mathbf{r}'_1, \mathbf{r}'_2, \ldots, \mathbf{r}'_{N-1}, \ldots, \mathbf{r}'_{N-1}, \rho, \zeta;\, r|n\rangle\, \langle r|v_n\rangle, \quad (11.51)$$

In this approximation, one can introduce the electronic transition moment,

$$\begin{aligned}\mathcal{R}_{n'n}(r) &= \langle n'|T^{(q)}_{k,\Omega-\Omega'}(\mathbf{r}'_1\mathbf{r}'_2\ldots\mathbf{r}'_{N-1}\rho\ \zeta;\ r)|n\rangle \\ &= b_0 + b_1 r + b_2 r^2 + \ldots,\end{aligned} \tag{11.52}$$

its vibrational matrix elements,

$$\langle v_{n'}|\mathcal{R}_{n'n}(r)|v_n\rangle = b_0\langle v_{n'}|v_n\rangle + b_1\langle v_{n'}|r|v_n\rangle + b_2\langle v_{n'}|r^2|v_n\rangle + \ldots. \tag{11.53}$$

and these in turn lead to the Franck–Condon factors and r-centroids,

$$|\langle v_{n'}|\mathcal{R}_{n'n}(r)|v_n\rangle|^2 = q(v_{n'},\ v_n)[b_0 + b_1\ \bar{r}^{(1)}(v_{n'},\ v_n) + b_2\ \bar{r}^{(2)}(v_{n'},\ v_n) + \ldots]^2, \tag{11.54}$$

$$q(v_{n'},\ v_n) = \langle v_{n'}\ |v_n\rangle^2, \tag{11.55}$$

$$\bar{r}^{(k)}(v_{n'},\ v_n) = \frac{\langle v_{n'}v\ |r^k|v_n\rangle}{\langle v_{n'}\ |v_n\rangle}. \tag{11.56}$$

In summary, the Born–Oppenheimer approximation gives

$$S(n'v_{n'}J',\ nv_nJ) = q(v_{n'},\ v_n)[b_0 + b_1\ \bar{r}^{(1)}(v_{n'},\ v_n) + b_2\ \bar{r}^{(2)}(v_{n'},\ v_n) + \ldots]\ S(J',\ J') \tag{11.57}$$

for the diatomic electronic–vibrational line strength. Almost invariably, the accuracy of line position measurement exceeds the accuracy of even relative intensity measurement by several orders of magnitude. The calculation of accurate line positions using the Born–Oppenheimer approximation requires that the vibrational eigenfunction $\langle r\ |v\rangle$ be expressed as a large sum of Born–Oppenheimer vibrational eigenfunctions $\langle r\ |v_n\rangle$,

$$\langle r\ |v\rangle = \sum_{v_n}\langle r\ |v_n\rangle\ \langle v_n\ |v\rangle. \tag{11.58}$$

The sum over v_n produces large Hamiltonian matrices whose dimensions are reduced to manageable size by Van Vleck transformations or other comparable computations, and, when successful, the eigenvalues of the smaller matrices can accurately reproduce experimentally determined terms Typically, only the simple Born–Oppenheimer approximation result (11.57) is required for diatomic intensity predictions.

If one's definition of an ugly number is a number too small in magnitude to be represented in standard single precision floating point digital format, then the line strength can be fairly described as an ugly number. It is normally expressed in units of $(a_0e)^2 = 7.1882479 \times 10^{-59}$ m^2 Coul^2 where a_0 is the Bohr radius and e is the charge of the electron.

Reference

[1] Rose M E 1995 *Elementary Theory of Angular Momentum* (Mineola, NY: Dover)

IOP Publishing

Quantum Mechanics of the Diatomic Molecule with Applications

Christian G Parigger and James O Hornkohl

Chapter 12

Using the Morse potential in diatomic spectroscopy

12.1 Introduction

A diatomic basis state has a potential energy function $V(r)$. The Morse potential is a simple mathematical formula one can use to approximate $V(r)$. If the spectral line vacuum wavenumbers have been experimentally measured to high J values for all known bands of the band system in question, then the numerical RKR potential energy curves are much better approximations of $V(r)$'s for the upper and lower basis functions than the Morse potential. Using RKR $V(r)$'s one can compute the vibrational eigenfunctions, Franck–Condon factors, r-centroids, and tables of G_v and B_v versus v for comparison with the experimental tables of G_v and B_v used to compute the RKR potentials. However, the experimental record for the band system might be often incomplete. The Morse potential is particularly useful when one needs the quantities that are computable from vibrational eigenfunctions but the experimental record is too incomplete for computation of RKR potentials.

Only four variables, ω_e, $\omega_e x_e$, B_e, and the reduced nuclear mass μ, are required for computation of the Morse potential,

$$V^{(M)}(r) = D_e[1 - e^{-\beta(r-r_e)}]^2 \tag{12.1}$$

where r is the internuclear distance and r_e is the value of r at the minimum of $V^{(M)}(r)$. The dissociation energy D_e, measured from the minimum of the $V^{(M)}(r)$ potential energy well, is given by

$$D_e = \frac{\omega_e^2}{4\omega_e x_e}. \tag{12.2}$$

The parameter β (unrelated to the Euler angle β) is given in terms of the anharmonicity constant $\omega_e x_e$ and the reduced nuclear mass μ in amu (atomic mass units),

$$\beta = \sqrt{\frac{4\pi c\ (\mu/N_{\mathrm{A}})}{\hbar}\omega_e x_e} \tag{12.3}$$

in which N_{A} is Avogadro's number. The classical equilibrium internuclear distance r_e is given in terms of the constant B_e,

$$r_e = \sqrt{\frac{\hbar}{4\pi c\ (\mu/N_{\mathrm{A}})}\frac{1}{B_e}}. \tag{12.4}$$

The energy eigenvalues of the Morse oscillator are

$$E_v = \omega_e(v + 1/2) - \omega_e x_e(v + 1/2)^2. \tag{12.5}$$

The constant B_e is that from the semi-empirical formula for the rotational constant,

$$B_v = B_e - \alpha_e(v + 1/2) + \gamma_e(v + 1/2)^2 + \cdots. \tag{12.6}$$

If one chooses the values of ω_e and $\omega_e x_e$ to be those in the semi-empirical formula for vibrational term G_v,

$$\begin{aligned} G_v = \omega_e(v + 1/2) &- \omega_e x_e(v + 1/2)^2 \\ &+ \omega_e y_e(v + 1/2)^3 + \omega_e z_e(v ++ 1/2)^4, \end{aligned} \tag{12.7}$$

then one intuitively expects that the lower region of the Morse potential $V^{(M)}(r)$ will closely approximate the same lower region of the RKR potential. Computation of the RKR potential requires full experimentally determined tables of G_v and B_v versus v. If experiment has provided G_v and B_v for only the lowest values of the vibrational quantum number v, one can nevertheless compute a Morse potential which usefully approximates the lower region of the RKR potential. Because the eigenfunctions of the Morse oscillator are analytically known (see the following section), one can compute the Franck–Condon factors and r-centroids and, at least for the few lower vibrational states, proceed as if the RKR potential were known. Far from complete tables of G_v and B_v are a common occurrence, and thus the Morse potential is a useful addition to the toolbox of applied spectroscopy.

12.2 Morse eigenfunctions

The variable x,

$$x = ae^{-\beta(r-r_e)} \tag{12.8}$$

in which

$$a = \frac{\omega_e}{\omega_e x_e} \tag{12.9}$$

is a more convenient variable than the internuclear distance r for expressing the Morse potential,

$$V^{(M)}(x) = D_e\left(1 - \frac{x}{a}\right)^2 \tag{12.10}$$

and its energy eigenfunction,

$$\psi_v^{(M)}(x) = \langle x|v\rangle_M = \mathcal{N}_v e^{-x/2} x^{\alpha_v/2} L_v^{(\alpha_v)}(x). \tag{12.11}$$

The normalization factor is

$$\mathcal{N}_v = \sqrt{\frac{\alpha_v \beta\ v!}{\Gamma(a - v)}} \tag{12.12}$$

in which the parameter α_v is given by

$$\alpha_v = a - 2v - 1 \tag{12.13}$$

and Γ is the gamma function.

The Laguerre polynomial $L_n^{(\alpha)}(x)$ is defined only for $\alpha > -1$. The requirement

$$\alpha_v > -1 \tag{12.14}$$

imposes the limit

$$v_{\max} < \frac{a}{2} \tag{12.15}$$

on the maximum value of the vibrational quantum number v for the Morse oscillator. That is, the Morse oscillator is bound for v less than $a/2$, but will dissociate if it is excited beyond the $v_{\max}$ bound state. For a diatomic potential energy function, this is significant improvement over the harmonic oscillator potential which never dissociates.

The Laguerre polynomials $L_v^{(\alpha_v)}(x)$ of the Morse eigenfunction are distinct from the usual Laguerre polynomials $L_n^{(\alpha)}(x)$ in that the parameter α in $L_n^{(\alpha)}(x)$ does not vary with the degree n of the polynomial, whereas α_v in $L_v^{(\alpha_v)}(x)$ does vary with the degree v. The mathematical manifestation of the variation of α_v with v is that $L_v^{(\alpha_v)}(x)$ does not satisfy a three-term recursion formula like that which $L_n^{(\alpha)}(x)$ with fixed α satisfies. This slightly complicates computation of the Morse eigenfunction and limits the practicality of one using a sum of the Morse eigenfunctions as a basis for expressing the eigenfunction of an arbitrary potential $V(r)$. The Lanczos basis is often expressed in terms of the spectroscopic quantities ω_e, $\omega_e x_e$, and r_e, but it uses $L_n^{(\alpha)}(x)$ with a fixed α.

A diatomic potential energy function should satisfy the limits

$$\lim_{r\to 0} V(r) \to \infty, \tag{12.16a}$$

$$\lim_{r\to\infty} V(r) = D_e, \tag{12.16b}$$

but the Morse potential function remains finite for negative fictitious negative values of the internuclear distance [1],

$$\lim_{r \to 0} V^{(M)}(r) \to \infty, \tag{12.17a}$$

$$\lim_{r \to 0} V^{(M)}(r) = D_e. \tag{12.17b}$$

This means that the equations for the Morse energy eigenvalues (12.5) and eigenfunctions (12.11) are approximations. However, for any diatomic molecule, the value of the Morse potential ar $r = 0$,

$$V^{(M)}(r = 0) = D_e e^{\beta r_e} \tag{12.18a}$$

$$= D_e e^{\sqrt{\omega_e x_e / Be}} \tag{12.18b}$$

is so large that equations (12.5) and (12.11) are effectively exact for all practical purposes in diatomic spectroscopy.

12.2.1 Computation of Morse eigenfunctions

Equation (12.11) is clearly simple, but its numerical evaluation is not entirely trivial. Computation of $L_v^{(\alpha_v)}(x)$ using the three-term recurrence formula

$$(n + 1)L_{n+1}^{(\alpha_v)}(x) = (2n + \alpha_v - x)L_n^{(\alpha_v)}(x) - (n + \alpha_v)L_{n-1}^{(\alpha_v)}(x) \tag{12.19}$$

means that one computes the unneeded $L_0^{(\alpha_v)}(x)$, $L_1^{(\alpha_v)}(x)$, ..., $L_{v-1}^{(\alpha_v)}(x)$ just to obtain the desired $L_v^{(\alpha_v)}(x)$. Given the computational power of even a modest digital computer, this wasted computation is likely of no practical consequence.

Of more significant nuisance are the values of $L_n^{(\alpha)}(x)$ large enough to cause numerical overflow. Shen *et al* [2] define 'generalized Laguerre functions (GLFs)',

$$\hat{\mathcal{L}}_n^{(\alpha)}(x) \equiv e^{-x/2} L_n^{(\alpha)}(x) \tag{12.20}$$

which follow the same recurrence formula as the standard $L_n^{(\alpha)}(x)$,

$$(n + 1)\hat{L}_{n+1}^{(\alpha_v)}(x) = (2n + \alpha_v - x)\hat{L}_n^{(\alpha_v)}(x) - (n + \alpha_v)\hat{L}_{n-1}^{(\alpha_v)}(x). \tag{12.21}$$

The two lowest degree $\hat{L}_n^{(\alpha_v)}(x)$ required to start the recursion are

$$\hat{L}_0^{(\alpha_v)}(x) = e^{-x/2}, \tag{12.22a}$$

$$\hat{L}_1^{(\alpha_v)}(x) = (\alpha + 1 - x)e^{-x/2}. \tag{12.22b}$$

Rewritten in terms of the GLFs, the Morse eigenfunction equation (12.11) reads

$$\psi_v^{(M)}(x) = \mathcal{N}_v x^{\alpha_v/2} \hat{L}_v^{(\alpha_v)}(x). \tag{12.23}$$

The FORTRAN-90 program MorseFCF evaluates eigenfunctions for the upper and lower vibrational states,

$$\psi_v^{(M)}(r) = e^{[\ln(\mathcal{N}_v)+\alpha_v \ln(x)]}\hat{L}_v^{(\alpha_v)}(x), \tag{12.24}$$

and uses the vibrational eigenfunctions to compute the Franck–Condon factors and the first three r-centroids. This program is particulary useful when experimental data for the band system are inadequate for accurate determination of the upper and lower potential energy functions. The subroutines in MorseFCF.f90 are potentially useful in other applications of Morse eigenfunctions. The source codes are listed in appendix I.

12.3 Morse eigenfunctions as a vibrational basis

Greenawalt and Dickinson [3] explored the use of Morse eigenfunctions as a basis for numerically expressing the solutions for a potential function $V(r)$ for which an analytic solution is not available. Dissociation of the Morse oscillator limits the maximum number of Morse basis function (equation (12.15)). In contrast, the extent of, for example, a harmonic oscillator basis is limited by the computer on which they are used.

The matrix elements of x, equation (12.8), evaluated in the Morse basis, equation (12.11), are

$$\langle n|x|n\rangle = a - 2n - 1 \tag{12.25a}$$

$$\langle n|x|m\rangle = (-)^{m-n}\frac{\mathcal{N}_n\mathcal{N}_m\Gamma(a - m)}{\beta n!}, \quad m \geqslant n \tag{12.25b}$$

$$\langle n|x^2|n\rangle = a(a - 2n - 1) \tag{12.26a}$$

$$\langle n|x^2|m\rangle = (-)^{m-n}\frac{\mathcal{N}_n\mathcal{N}_m\Gamma(a - m)}{\beta n!}[a(m - n + 1) - (m - n)(m + n + 1)] \tag{12.26b}$$

From these matrix elements and equation (12.10), one sees that matrix elements of the Morse potential in the Morse basis are given by

$$V_{nm}^{(M)} = D_e\left(1 - \frac{2\langle n|x|m\rangle}{a} + \frac{\langle n|x^2|m\rangle}{a^2}\right). \tag{12.27}$$

Since the diagonal Morse Hamiltonian matrix is known, equation (12.7), the kinetic energy matrix is simply the Hamiltonian matrix minus the potential matrix,

$$K_{nm}^{(M)} = G_n^{(M)}\delta_{nm} - V_{nm}^{(M)}. \tag{12.28}$$

The solutions for an unknown potential are found by computing the matrix elements of the unknown potential in the Morse basis, adding the Morse kinetic energy $K^{(M)}$ matrix, and diagonalization of the result.

Greenawalt and Dickinson [3] found that the Morse basis is useful for shallow potential wells. However, the matrix of x given in equations (12.25b) is a full symmetric matrix, not a tridiagonal matrix. This means that the eigenvalues of the x matrix are not the roots of an orthogonal polynomial satisfying a three-term recurrence formula, that the eigenvalues of x are not the arguments of a Gaussian integration. When numerical diagonalization of large matrices was made practical by advances in computer hardware and software in the 1960s, matrix solutions of the Schrödinger equation have become widespread. Early examples are [4–6], and [7].

Given the orthogonal polynomials of degree $n = 0$ to N having the weight function $w(x)$ over the domain $a \to b$, one can compute the matrix elements of x,

$$x_{nm} = \int_a^b w(x) P_n(x) x P_m(x)\, dx \tag{12.29}$$

A similarity transformation in which S is an orthogonal matrix (i.e. the inverse S^{-1} equals the transpose $\tilde{S}$) yields the eigenvalues λ_i of the x matrix,

$$\lambda_j = \sum_{n=0}^{N} \sum_{m=0}^{N} \tilde{S}_{jn} x_{nm} S_{mk}. \tag{12.30}$$

This equation can be inverted to give the matrix elements x_{nm} in terms of the eigenvalues λ_j,

$$x_{nm} = \sum_{i=0}^{N} \sum_{j=0}^{N} S_{ni} \lambda_j \delta_{ji} \tilde{S}_{jm} \tag{12.31a}$$

$$= \sum_{j=0}^{N} S_{nj} \lambda_j \tilde{S}_{jm} \tag{12.31b}$$

$$= \sum_{j=0}^{N} S_{nj} \lambda_j S_{mj} \tag{12.31c}$$

Comparing this last result with the conversion of the continuous integral equation (12.29) to a discrete sum using Gaussian integration,

$$x_{nm} = \sum_{j=0}^{N} w(\lambda_j) P_n(x_j) \lambda_j P_m(\lambda_j), \tag{12.32}$$

one sees that the two equations become equivalent if one makes the association

$$S_{nj} = \sqrt{w(\lambda_j)}\, P_n(\lambda_j). \tag{12.33}$$

The practical significance is that equation (12.32) yields exact values for all x_{nm} except x_{NN}.

References

[1] Ter Haar D 1946 *Phys. Rev.* **70** 222
[2] Shen J, Tang T and Wang L-L 2010 *Spectral Methods, Algorithms, Analysis and Applications* (Heidelberg: Springer)
[3] Greenawalt E M and Dickinson A S 1969 *J. Mol. Spectrosc.* **30** 427
[4] Chan S I and Stelman D 1963 *J. Chem Phys.* **39** 545
[5] Harris D O, Engerholm G G and Gwinn W D 1965 *J. Chem. Phys.* **43** 1515
[6] Zetik D F and Matsen F A 1967 *J. Mol. Spectrosc.* **24** 122
[7] Dickinson A S and Certain P R 1968 *J. Chem. Phys.* **49** 4209

Part II

Selected applications of diatomic spectroscopy

Chapter 13

Introduction to applications of diatomic spectroscopy

Diatomic spectroscopy encompasses the analysis of the interaction of electromagnetic radiation with diatomic molecules. Applications discussed in this book primarily explore molecular characteristics in and near the visible region of the electromagnetic spectrum, but extending to the near-UV and near-IR in the range of 200–800 nm. A variety of experimental studies are designed first for identification of molecular signatures or 'fingerprints' in measured spectra, and second, for rigorous analysis of recorded data. Of course, analytical chemistry explores opportunities for the establishment of well-defined diagnostic protocols for diatomic molecules.

Accurate predictions of molecular spectra are essential for analysis. Ultimately, computed diatomic spectra are expected to be readily available together with formalized comparison models for diagnosis purposes. In atomic spectroscopy, species identification relies on occurrences of specific lines and sequences of lines. For example, measurement of a line near 486.1 nm may not be enough for conclusively identifying hydrogen, but usually requires positive checks near 656.2 nm, and equally near 434.0 nm. With atomic signatures of the hydrogen alpha-, beta-, and gamma-lines, one can conclude with confidence the presence of hydrogen due to the occurrence of the first three members of the Balmer series lines in the visible portion of the electromagnetic spectrum. Conversely, as one looks for sodium, one would expect signatures of the Na D_1 and D_2 lines near 589.6 and 589.0 nm. Molecular spectroscopy diagnosis works in an analogous manner as one aims to identify presence of molecules in survey spectra. Fingerprints of molecular spectra measured with spectral resolutions of the order of 0. 1 nm include location of the band head, shading of the $\Delta v = 0$ sequence, and identification of the progression. For example, molecular OH spectra indicate a well-developed, $\Delta v = 0$ band head near 306 nm, are shaded toward the red, and show a $\Delta v = 1$ band head near 281 nm. Overlapping or interfering

doi:10.1088/978-0-7503-1989-8ch13

spectral regions may require detailed identification work using well-documented tables [1]. In turn, presence of carbon C_2 Swan spectra in recorded plasma emission data implies the presence of radiating C_2 diatomic molecules.

Ideally, diatomic molecular spectra should be available in a database or should be easily computed. To this end, recent publications [2, 3] describe two programs that accomplish prediction of signatures from line strength data for selected transitions of AlO, C_2, CN, and TiO diatomic molecules and fitting of measured spectra, namely, Boltzmann equilibrium spectrum computation (BESP) and Nelder–Mead temperature (NMT) programs. However, several other diatomic molecules have been of interest in laser spectroscopy work by the authors, e.g. OH to name but one other example. Figures 13.1–13.7 show selected synthetic spectra for a few diatomic molecules that have been of continued interest: OH, AlO, and C_2 Swan band overview (progression) and selected sequences $\Delta\nu = 0, \pm 1, \pm 2$ of the C_2 Swan bands, and $\Delta\nu = 0, \pm 1$ of CN.

Applications of the computed line-strengths include analysis of molecular signatures in laser-induced breakdown spectroscopy, combustion studies, and analysis of stellar astrophysical spectra. Laser-induced breakdown spectroscopy (LIBS) aims to diagnose elemental composition following generation of a laser spark using analytical chemistry methods analogous to discharge spark spectrometry. A mini-review of diatomic LIBS is presented in appendix H.

Molecular signatures are recognizable in the emitted plasma radiation within the first microsecond after laser-induced optical breakdown, especially when utilizing nominal femto- to several tens of pico-second pulses for generation of laser plasma.

For nominal, 5–10 ns radiation, decaying plasma characteristically reveals relatively large background radiation that masks molecular emissions, but molecular spectra are well-developed in the plume and for time delays of the order of 10–100 µs after initiation of optical breakdown. Combustion studies obviously focus on the

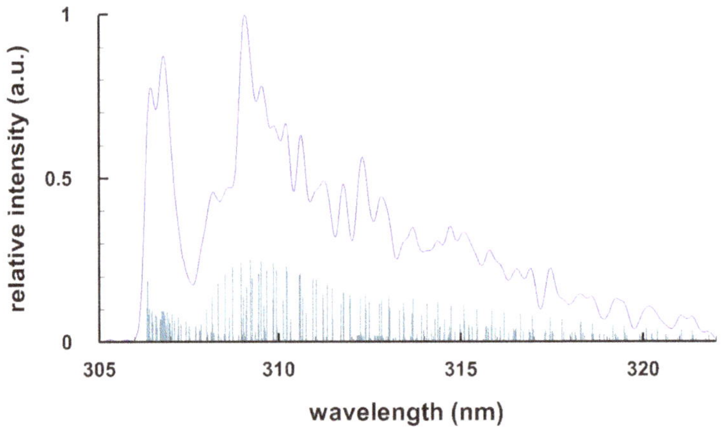

Figure 13.1. Computed spectrum of the $A^2\Sigma \rightarrow X^2\Pi$ UV band of OH, $T = 4$ kK, (top) spectral resolutions of $\Delta\lambda = 0.32$ nm ($\Delta\tilde{\nu} = 32$ cm^{-1}) and (bottom) idealized resolution for the stick spectrum $\Delta\lambda = 0.002$ nm ($\Delta\tilde{\nu} = 0.2$ cm^{-1}) of the $\Delta\nu = 0$ sequence [2].

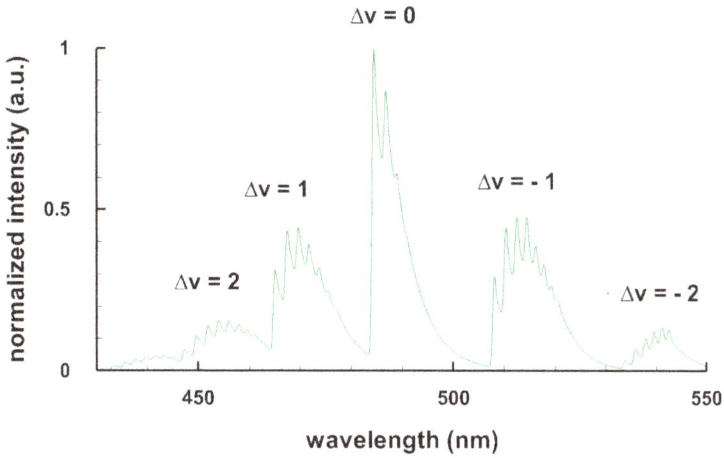

Figure 13.2. Computed spectrum of the AlO $B^2\Sigma^+ \rightarrow X^2\Sigma^+$ band, T = 4 kK, for a spectral resolution of $\Delta\lambda = 0.78$ nm ($\Delta\tilde{\nu} = 32$ cm^{-1}). Indicated are the $\Delta\nu = 0, \pm 1, \pm 2$ sequences of the AlO progression. Reprinted from [2, 3], copyright (2015), with permission from Elsevier.

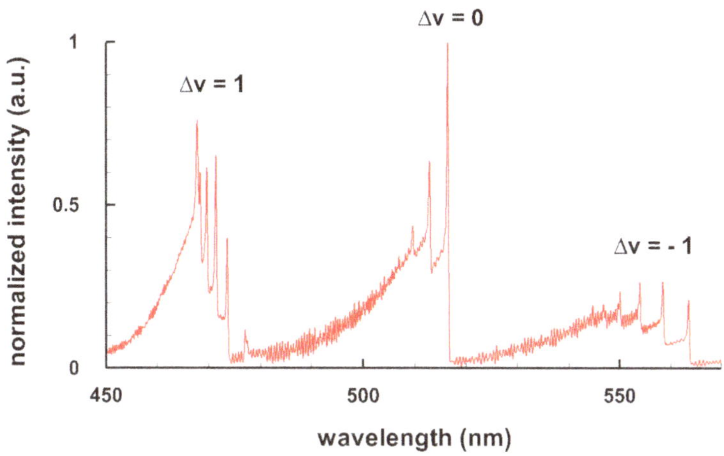

Figure 13.3. Computed spectrum of C_2 Swan $d^3\Pi_g \rightarrow a^3\Pi_u$ band progression, T = 8 kK, for a spectral resolution of $\Delta\lambda = 0.32$ nm ($\Delta\tilde{\nu}$ = 12 cm^{-1}). Indicated are the $\Delta\nu = 0, \pm 1$ sequences of the C_2 progression. Reprinted from [2, 3], copyright (2015), with permission from Elsevier.

presence and dynamics of molecules that would indicate combustion, viz. OH in hydrocarbon combustion or NH in ammonia combustion. Stellar astrophysical spectra from white dwarfs at a temperature of the order of 30 kK indicate primarily atomic hydrogen spectra in absorption. As white dwarfs cool to temperatures of the order of 8 kK, molecular spectra are frequently recorded in absorption. The establishment of a connection between astrophysical spectra measured in absorption and laboratory emission spectra measured in the laboratory requires correction for bound-free radiation, but otherwise the established molecular line strengths for molecules such as C_2 and CN are well-suited for astrophysical analysis.

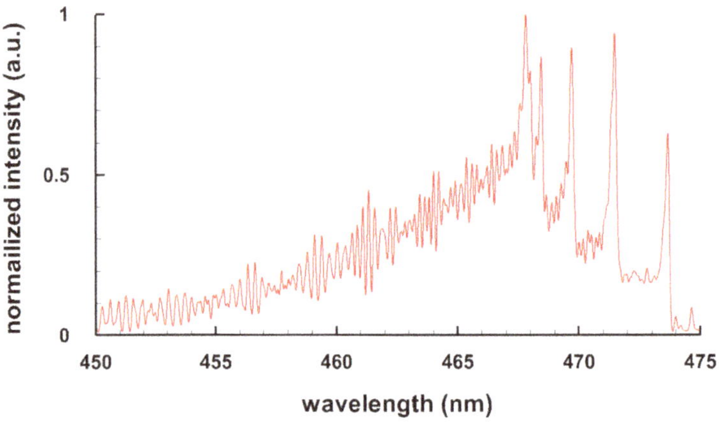

Figure 13.4. C_2 Swan $d^3\Pi_g \rightarrow a^3\Pi_u$ band $\Delta\nu = -1$ sequence, $T = 8$ kK, $\Delta\lambda = 0.13$ nm ($\Delta\tilde{\nu} = 6$ cm^{-1}) [2].

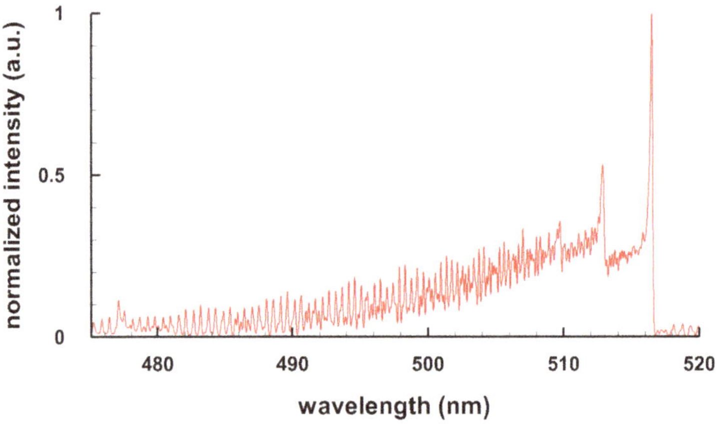

Figure 13.5. C_2 Swan $d^3\Pi_g \rightarrow a^3\Pi_u$ band $\Delta\nu = 0$ sequence, $T = 8$ kK, $\Delta\lambda = 0.15$ nm ($\Delta\tilde{\nu} = 6$ cm^{-1}) [2].

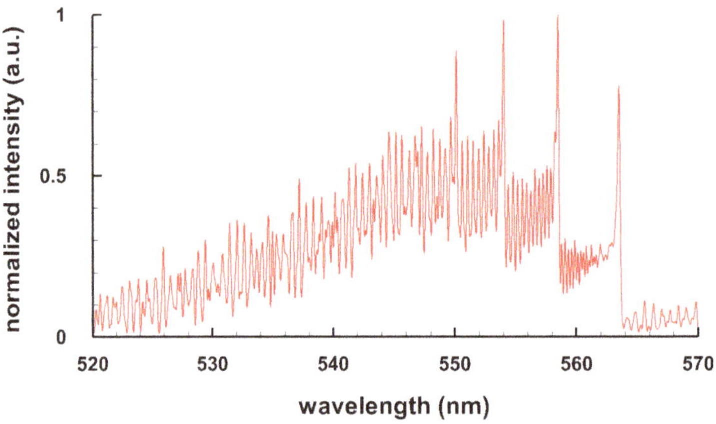

Figure 13.6. C_2 Swan $d^3\Pi_g \rightarrow a^3\Pi_u$ band $\Delta\nu = +1$ sequence, $T = 8$ kK, $\Delta\lambda = 0.18$ nm ($\Delta\tilde{\nu} = 6$ cm^{-1}) [2].

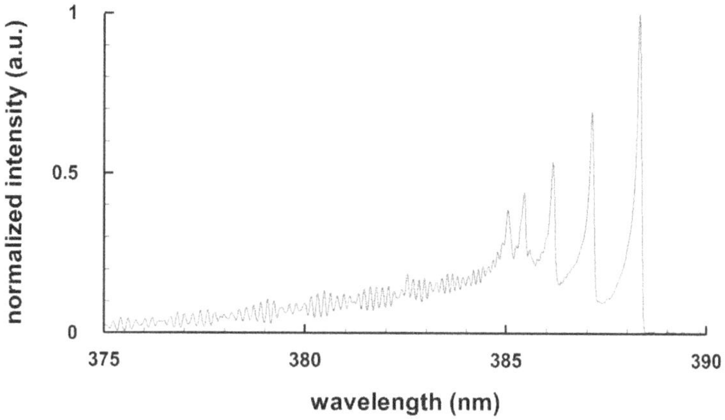

Figure 13.7. Computed spectrum of the CN violet $B^2\Sigma^+ \rightarrow X^2\Sigma^+$ band, $T = 8$ kK, for a spectral resolution of $\Delta\lambda = 0.09$ nm ($\Delta\tilde{\nu} = 6\ \mathrm{cm}^{-1}$) of the $\Delta\nu = 0$ sequence. Reprinted from [2, 3], copyright (2015), with permission from Elsevier.

Several laboratory measurements are designed to validate the data sets developed with the theory communicate in part I of this book for selected molecular transitions, equally, the data sets for several molecules are utilized in the analysis of the plasma condition. Moreover, data sets are utilized as well in laser-induced fluorescence measurements in combustion studies, yet the primary application comprises analysis of the laser-plasma composition that will serve as diagnosis of the targets that may be gases, liquids, or solids, including nano-particles.

Typical experimental arrangements are communicated in the next chapter followed by several chapters on diatomic molecular spectroscopy.

References

[1] Wallace L 1962 *Astrophys. J. Suppl. S.* **7** 165

[2] Parigger C G and Hornkohl J O 2010 *Int. Rev. At. Mol. Phys.* **1** 25

[3] Parigger C G, Woods A C, Surmick D M, Gautam G, Witte M J and Hornkohl J O 2015 *Spectrochim. Acta Part B: At. Spectrosc.* **107** 132

Chapter 14

Experimental arrangement for laser-plasma diagnosis

Experimental investigations of diatomic molecular emission spectra utilize laboratory, table-top scale laser-, spectrometer-, and detector equipment. Laser-plasma is generated by focusing pulsed radiation to an irradiance sufficient to achieve optical breakdown at the target. The irradiance threshold for standard ambient temperature and pressure (SATP) dry air is of the order of 0.3 TW cm^{-2} that can be accomplished with $f/10$ focusing of 10 ns, 100 mJ/pulse, IR radiation at 1.064 µm radiation from a Q-switched Nd:YAG device. For example, the ratio of a single 10 cm focal-length lens and a 1 cm beam diameter at the lens amounts to 10, labeled as $f/10$ focusing. The spot-size diameter is approximately the product of wavelength and f-number; therefore, the cross-section focal area is of the order of 10^{-6} cm^2. The peak power of a 100 mJ, 10 ns pulse is 10 MW, consequently, the peak irradiance in focus is of the order of 10 TW cm^{-2}. However, detailed calculation that include consideration of physical optics and primarily spherical aberration correction would reveal a multiple-peak distribution in the focal volume. The focal volume is defined by the Rayleigh-length that extends from either side of focus to an area cross-section twice that of the focal area. Typical results are communicated in recent laser-induced breakdown spectroscopy (LIBS) [1, 2].

Recent experimental arrangement for laboratory measurements of laser-induced plasma [3] includes a Q-switched Nd:YAG laser device that generates 6 ns laser pulses with an energy of 850 mJ per pulse at the fundamental wavelength of 1064nm. Radiation is tightly focused to a spot size of the order of 10 µm to achieve a peak irradiance above 1 TW cm^{-2}, or well above optical breakdown in SATP laboratory air and gaseous hydrogen at a pressure of the order of one atmosphere. The emanating light from the micro-plasma is recorded with a Czerny-Turner type spectrometer that shows a resolution of 0.02 nm when using a 3600 grooves/mm

doi:10.1088/978-0-7503-1989-8ch14

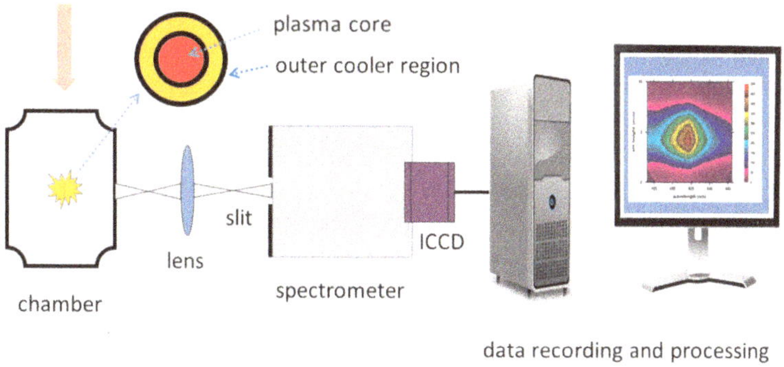

Figure 14.1. Optical breakdown plasma is generated by focusing the laser beam from the top and parallel to the slit. Time-resolved 1:1 images are recorded with the ICCD [3].

holographic grating and an intensified diode array or two-dimensional detector such as an intensified charge-coupled detector (ICCD) for temporally- and spatially-resolved spectroscopy. Figure 14.1 illustrates the experimental schematic that is employed for atomic- and molecular-emission spectroscopy following optical breakdown in gases.

Several variants of the experimental arrangement are employed. For example, a cell is not required for studies of air breakdown, laser-ablation of solids and liquids or droplets, or nano-particles. Also, for studies of laser-induced ignition of combustible mixtures specially designed burners with capabilities of generating a shroud flow are utilized. Other variants of experiments discussed in this book include different laser-device for plasma generation to include a pico-second excimer laser operating at a wavelength 0.308 μm [4] and a CO_2 laser operating at 10.6 μm [5]. A scanning dye laser is employed for planar laser-induced fluorescence experiments. Various details on laser spectroscopy and high peak power laser–matter interactions are elaborated in recent books [6–9]. Wavelength and sensitivity corrections are important for comparison of recorded and computed data due to the wavelength range of molecular spectra, especially when considering rotation–vibration spectra sequences and progressions. For a recent summary on LIBS that shows 6–8 kK excitation temperature molecular emission spectra see Miziolek *et al* [10] and Jagdish and Thakur [11].

Detailed equipment lists are communicated in the handbook by Cremers and Radziemski [12] for laser-induced breakdown spectroscopy. The equipment list and performance parameters are of course applicable for molecular plasma-emission spectroscopy. Recent laboratory methods are communicated in a video [13] and are routinely applied in the Center for Laser Application (CLA) at the University of Tennessee Space Institute (UTSI). The text associated with [13] conveys a detailed list of equipment for atomic and molecular spectroscopy.

The analysis of the measured spectra uses the concepts and methods discussed in part I of this book. For isolated molecular transitions, the Franck–Condon factors are found by solving the radial Schrödinger equation numerically, and the Hönl–London strengths are obtained by numerical diagonalization of the rotational and fine-structure Hamiltonian [14]. The computation of a diatomic spectrum is accomplished by specifying the temperature and spectral resolution. In turn, non-linear fitting routines can be employed infer the spectroscopic temperature, spectral resolution for the fitting, and background contributions due to blackbody radiation or contributions from other species [15]. Of interest are diatomic molecules such as AlO, C_2, CH, CN, CrF, N_2^+ 1^{st}Neg, N_2 1^{st}Pos, N_2 2^{nd}Pos, NH, NO, OH. Accurate line-strength files are in use for spectroscopic analysis with the so-called Boltzmann Equilibrium Spectrum Program (BESP) [16]. The BESP is used, for example, in the analysis of laser-induced optical breakdown and laser-induced fluorescence. The computed or synthetic spectra have been instrumental in the analysis of molecular laser-induced optical breakdown, and equally, for experimental check-points of the computed data base for diatomic molecular spectroscopy, see, for example, [4, 5, 17–22].

References

[1] Parigger C G 2006 Laser-induced breakdown in gases: experiments and simulation *Laser-Induced Breakdown Spectroscopy(LIBS), Fundamentals and Applications* ed A W Miziolek, V Palleschi and I Schechter (New York: Cambridge University Press) ch 4

[2] Parigger C G, Surmick D M, Helstern C M, Gautam G and Bol'shakov A A 2019 Molecular laser-induced breakdown spectroscopy *Laser-Induced Breakdown Spectroscopy* ed J P Singh and S N Thakur (New York: Elsevier) ch 7

[3] Parigger C G, Drake K A, Helstern C M and Gautam G 2018 *Atoms* **6** 36

[4] Hornkohl J O, Parigger C G and Lewis J W L 1991 *J. Quant. Spectrosc. Radiat. Transfer* **46** 405

[5] Parigger C G, Hornkohl J O and Nemes L 2010 *Int. J. Spectrosc.* **2010** 159382

[6] Kunze H-J 2009 *Introduction to Plasma Spectroscopy* (Heidelberg: Springer)

[7] Bauer D and Mulser P 2010 *High Power Laser-Matter Interaction* (Heidelberg: Springer)

[8] Hertel I V and Schulz C-P 2015 *Atoms, Molecules and Optical Physics 1, Atoms and Spectroscopy* (Heidelberg: Springer)

[9] Hertel I V and Schulz C-P 2015 *Atoms, Molecules and Optical Physics 2, Molecules and Photons—Spectroscopy and Collisions* (Heidelberg: Springer)

[10] Miziolek A W, Palleschi V and Schechter I (ed) 2006 *Laser-Induced Breakdown Spectroscopy (LIBS)—Fundamentals and Applications* (New York: Cambridge University Press)

[11] Singh J P and Thakur S N (ed) 2019 *Laser-Induced Breakdown Spectroscopy* (New York: Elsevier)

[12] Cremers D E and Radziemski L J 2006 *Handbook of Laser-Induced Breakdown Spectroscopy* (New York: Wiley)

[13] Parigger C G, Woods A C, Witte M J, Swafford L D and Surmick D M 2014 *J. Vis. Exp.* **84** 51250

[14] Nemes L, Hornkohl J O and Parigger C G 2005 *Appl. Opt.* **44** 3686

[15] Parigger C G, Woods A C, Surmick D M, Gautam G, Witte M J and Hornkohl J O 2015 *Spectrochim. Acta Part B: At. Spectrosc.* **107** 132

[16] Hornkohl J O and Parigger C G 1996 Boltzmann equilibrium spectrum program (BESP) (accessed January 19, 2016) http://view.utsi.edu/besp/

[17] Parigger C G, Plemmons D H, Hornkohl J O and Lewis J W L 1994 *J. Quant. Spectrosc. Radiat. Transfer* **52** 707

[18] Hornkohl J O, Nemes L and Parigger C G 2009 Spectroscopy, dynamics and molecular theory of carbon plasmas and vapors *Advances in the Understanding of the Most Complex High-Temperature Elemental System* ed L Nemes and S Irle (Singapore: World Scientific) ch 4
[19] Nemes L, Keszler A M, Hornkohl J O and Parigger C G 2005 *Appl. Opt.* **44** 3661
[20] Dors I G, Parigger C G and Lewis J W L 1998 *Opt. Lett.* **23** 1778
[21] Parigger C G, Guan G and Hornkohl J O 2003 *Appl. Opt.* **42** 5986
[22] Woods A C, Parigger C G and Hornkohl J O 2012 *Opt. Lett.* **37** 5139

Chapter 15

Cyanide, CN

Diagnosis of cyanide is important in a variety of fields of science. CN readily occurs as a result of recombination in laser-induced plasma in air. The measured signal strength is largely increased over that in air when 1:1 mole ratio mixtures of CO_2:N_2 are investigated [1, 2]. This chapter also summarizes recent research communicated at various national and international conferences.

The mole ratio and density of CN can be predicted for thermodynamic equilibrium using the so-called chemical equilibrium and applications (CEA) code [3, 4]. During the plasma expansion in the first microsecond after initiation of optical breakdown, shockwave phenomena affect the distribution of atomic and molecular species. Cyanide can be measured within the first 0.1 μs, but is well-developed after several dozens of microseconds. Diatomic molecular spectra of CN with excitation temperature in the range of 6–9 kK are also routinely measured in hydrocarbon combustion in air, the fuel providing carbon and the air providing nitrogen for the formation of CN due to combustion or due to recombination radiation in laser-induced breakdown plasma.

15.1 Analysis of CO_2 laser-plasma

Experimental investigations for measurements of CN utilize a standard laser-induced breakdown spectroscopy (LIBS) arrangement: laser device, spectrometer, detector-system, and a sample. For studies that use a CO_2 laser as the radiation source [5], experiment details are as follows: the CO_2 laser is the transversely excited atmospheric-pressure(TEA) model TEA-820 Lumonics, the spectrometer is the 1/2 meter model 500 SpectraPro Acton Research Corporation, and the detector is the intensified model 1460 EG&G Princeton Applied Research detector/controller Optical Multichannel Analyzer (OMA). The laser plasma is first generated for alignment purposes and initial diagnostic studies in laboratory air using *f*/4 focusing.

doi:10.1088/978-0-7503-1989-8ch15

The plasma is then generated in CO_2 gas that expands into the focal volume at a rate of 10 L/m into the laboratory. Survey spectra from laser-induced optical breakdown (LIOB) in CO gas are reported here.

Typical operating parameters of our CO_2 TEA laser are (i) a pulse width of 70–100 ns full-width at half-maximum, or 10× longer than Nd:YAG pulses, (ii) a pulse energy of 2000–3000 mJ, or typically 10–100× larger than nominal Nd:YAG laser devices, and (iii) a wavelength of 10.6 μm, or 10× longer than the 1.064 μm fundamental Nd:YAG laser radiation.

Figure 15.1 shows the recorded and fitted spectrum of the CN $\Delta v = 0$ sequence of the CN $B^2\Sigma^+ \longrightarrow X^2\Sigma^+$ violet system. The spectral resolution amounted to 0.12 nm when using the 2400 grooves/mm grating. Particularly strong CN emissions occur and dominate the molecular emissions in the visible region when using CO_2 flow expanding into laboratory air. We used a gate width of 20 μs and a time delay of 20 μs. The inferred temperature amounts to 7.22 kK, using our fitting routines for diatomic spectra [1, 5–7]. Figure 15.1 also shows an inset near the wavelength region of 4–4 band of CN that we excluded in the fitting routine. The additional spectroscopic feature indicates presence of the C_2 Deslandres D'Azambuja system [8].

15.2 Analysis of CN in Nd:YAG laser-plasma

Various applications including LIBS utilize Q-switched Nd:YAG laser pulses to generate plasma [1]. Figure 15.2 illustrates measured and fitted spectra of the CN violet system recorded with a spectral resolution of $\Delta\lambda = 0.03$ nm (2 cm^{-1}). Individual band heads of the Δv = 0 sequence are clearly discernible. The fitting of the experimental spectrum utilizes the line-strength data for CN and the NMT program as elaborated in the supplement of [9]. The computed spectra are based on the provided line strength data, and the fitting routine uses non-linear fitting based on simplices in the Nelder–Mead algorithm [10, 11].

The Nelder–Mead method [11] minimizes the difference between a fitting function and a set of data by iteratively adjusting the parameters of the fitting function. The method does not require that the derivative of the fitting function be known. For a

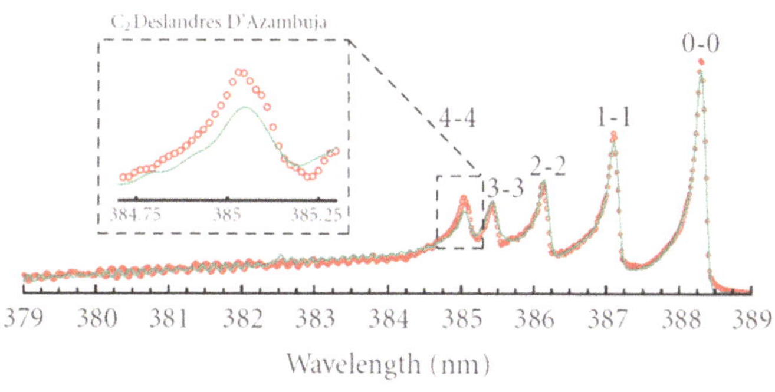

Figure 15.1. Measured (circles) and fitted (solid line) spectrum of the CN Δv = 0 sequence of the CN $B^2\Sigma^+ \longrightarrow X^2\Sigma^+$ violet system. The inset indicates emission from the C_2 Deslandres D'Azambuja system [5].

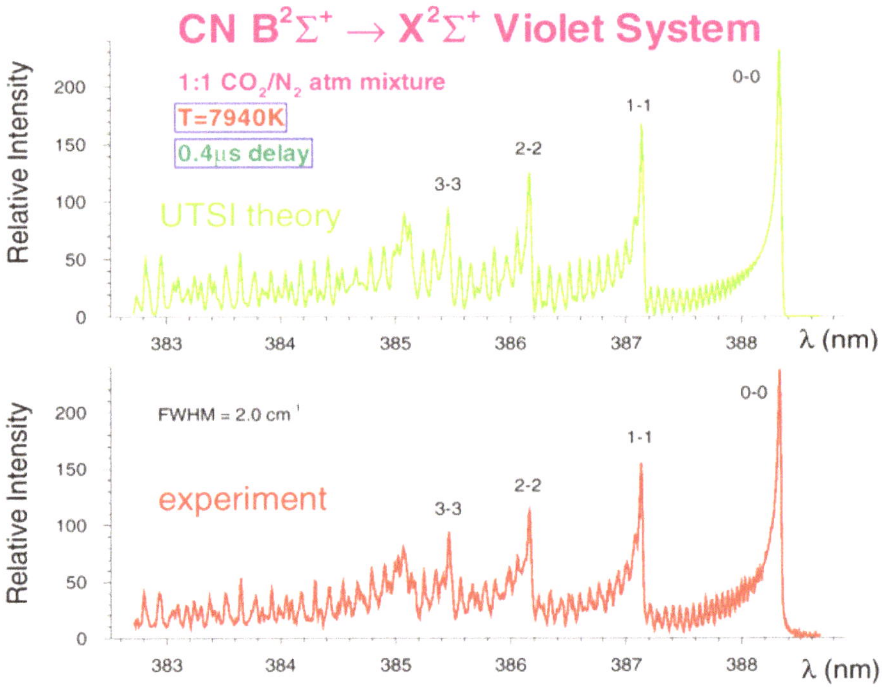

Figure 15.2. Measured (bottom) and fitted (top) spectrum of the CN $\Delta v = 0$ sequence of the CN $B^2\Sigma^+ \longrightarrow X^2\Sigma^+$ violet system. Reprinted from [1], copyright (1991), with permission from Elsevier.

set of N parameters, the Nelder–Mead method begins with $N + 1$ trial values of the parameters and $N + 1$ values of the difference between the fitting function and the data, and the method operates by moving away from the maximum difference. Each of the $N + 1$ trial values of the parameters and the $N + 1$ function differences is called a vertex. Each vertex is a point in an N-dimensional space called a simplex. In essence, the volume of the simplex is minimized, thus the name *downhill simplex*. Figure 15.3 displays measured-, fitted-, and difference-spectra at a spectral resolution of 0.093 nm.

Alternative programs that allow analysis of measured diatomic spectra include the programs LIFBASE [12] and PGOPHER [13]. The former appears to work well for lower excitation temperatures, and the latter has enjoyed significant updates and improvements for CN. The current (March 2019) pgo-file for the BAX CN transitions [14] delivers results that are consistent with the NMT fitting in terms of inferred temperature and fitted spectral resolution, see figure 15.3. Figure 15.4 compares experimental and computed spectra, but fitting algorithms in PGOPHER utilize vacuum wavenumbers, therefore requiring conversion from air wavelengths [15] in the experiment. Moreover, in the PGOPHER program relative intensities are used, in other words, computation of partition functions is avoided for plasma excitation temperatures of the order of 8 kK. The spectral resolution for PGOPHER

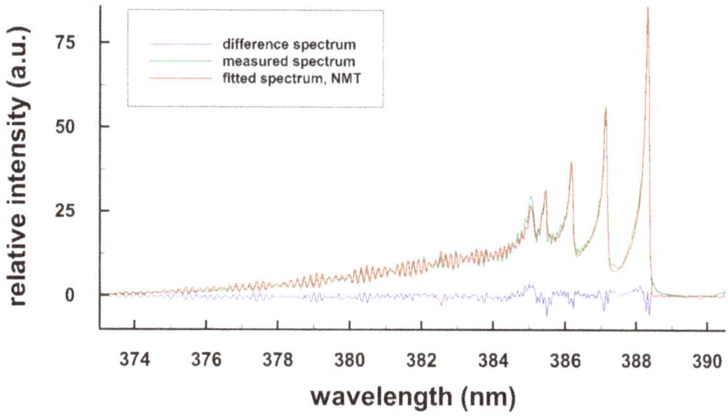

Figure 15.3. Measured, fitted, and difference spectra of the CN Δv = 0 sequence of the CN $B^2\Sigma^+ \longrightarrow X^2\Sigma^+$ violet system at a temperatire of $T = 6.84$ kK. Reprinted from [1, 9], copyright (1991), with permission from Elsevier.

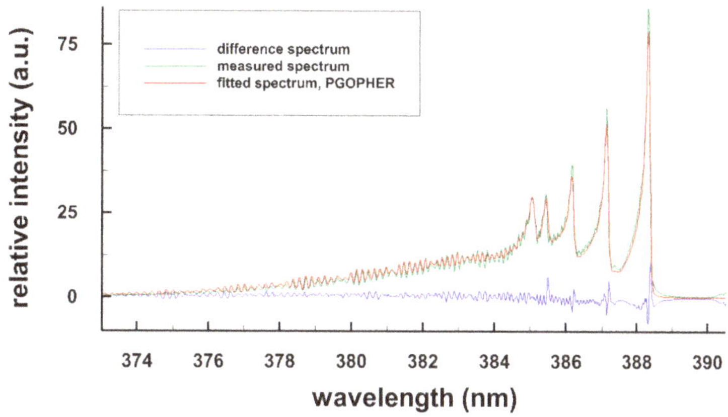

Figure 15.4. Results of fitting the experimental data of figure 15.3 with the PGOPHER program. Difference, measured, and fitted spectra for the CN Δv = 0 sequence of the CN $B^2\Sigma^+ \longrightarrow X^2\Sigma^+$ violet system yield a temperature of T = 6.83 kK. Reprinted from [1], copyright (1991), with permission from Elsevier.

fitting (version 10.1 of December 2018) is accomplished with 5.2 cm^{-1} Gaussian and 0.6 cm^{-1} Lorentzian widths.

15.3 Spatially and temporally resolved CN spectra

Recent experimental investigations explore the CN violet system using an intensified charge-coupled detector (ICCD) that is arranged to spatially resolve spectra along the slit dimension [2]. Laser-induced optical breakdown is generated by propagating the IR laser-beam from the top and parallel to the spectrometer slit, see the experimental arrangement in figure 14.1 in chapter 14. Initial experimental studies address details of the focal volume intensity distribution and effects on optical breakdown in laboratory air at standard ambient pressure and temperature (SATP).

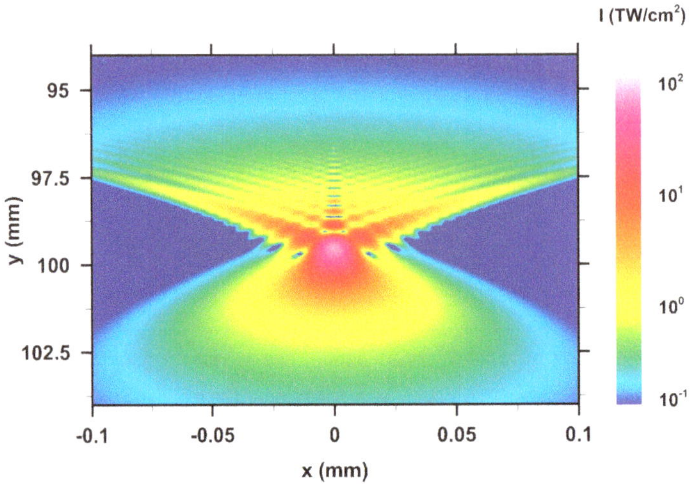

Figure 15.5. Computed focal distribution for *f*/10, 1064 nm focusing from the top [20].

15.3.1 Laser-beam focusing

A laser beam with 850 mJ/pulse is employed for plasma generation in SATP laboratory air, and another synchronized beam of 13 ns duration intersects the plasma to obtain shadow-graphs on a screen. A standard CCD is utilized for the recording of subsequent shadow-graphs. The focal distribution is obtained by solving diffraction integrals for a beam from the laser device that is expanded to a diameter of 10 mm. A fused silica, 100 mm, plano-convex lens (Thorlabs model LA4545) focuses the laser beam with *f*/10 optics. The *f*-number (*f*/#) is computed as the ratio of beam focal length and the beam diameter at the lens. Figure 15.5 illustrates computed distributions including aberrations [2, 16] with an aspect ratio of 50:1 for the Thorlabs LA4545 lens for focusing with *f*/10 optics.

The peak irradiance distributions are computed for 850 mJ, 6 ns, 1064 nm radiation. Tighter *f*/5 focusing for a perfect lens would show about one order of magnitude (or by a factor of 2^3) smaller focal volume than that obtained for *f*/10 focusing. The optical breakdown threshold is 0.28 TW cm^{-2} [17, 18] at 1 atm in dry air and for 1064 nm radiation. The yellow pseudo-color range near 1 TW cm^{-2} (10^0 in the figure) indicates an irradiance that is a factor of 3–4 higher than optical breakdown threshold.

Several regions along the optical axis indicate above-threshold irradiance as the laser-beam reaches focus near 100 mm. Consequently, optical breakdown will occur at multiple spots as indicated in the figure with subsequent absorption of the remainder of the 6 ns pulse. As the energy/pulse is reduced to about 8.5 mJ/pulse, optical breakdown would occur at focus causing a single spherical expansion.

The irradiance along the optical axis ($x = 0$) exhibits undulations above and below the perfect, aberration-free lens. Figure 15.6 shows optical-axis variations of the *f*/10 with aberrations. There are about ten maxima and minima for an irradiance larger than the clean air optical breakdown threshold of 0.28 TW cm^{-2} for clean air.

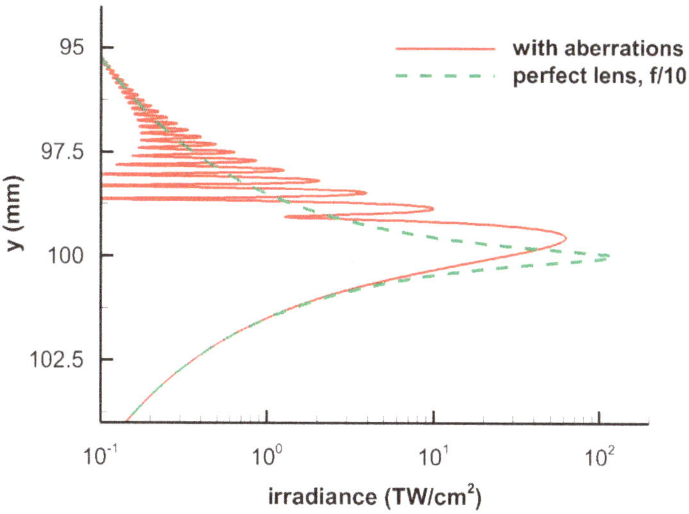

Figure 15.6. Irradiance along the optical axis for *f*/10 focusing.

15.3.2 Shadowgraphs

Optical breakdown visualizations utilize the shadow-graph method, or visualization based on the second derivative of the index of refraction. The shadow-graphs are recorded using a second laser beam that is externally synchronized to the pulse that generates optical breakdown [19]. Figure 15.7 illustrates optical breakdown for 0.05, 0.6, 1.5, and 3 μs time delays. The representative images are taken from a set of 124 recorded shadow-graphs from different breakdown events, in other words, figure 15.7 is put together from different experiments. The background shows the particular mode structure of the second beam used as source for the shadow graph. Clearly, there are multiple breakdown spots in the 0.05 μs image, and each of these spots indicates a spherically symmetric expansion. At threshold irradiance, only one laser-spot would be measured. The deviation from spherical expansion is indicated by the forward cone toward the bottom of the images, nicely developed in the 0.06 μs image.

15.3.3 Raw CN spectra

Figure 15.8 displays raw data obtained in the spectral range of the CN $\Delta v = 0$ sequence and illustrate the CN distribution along the plasma that is generated by focusing the beam parallel to the spectrometer slit. Of particular interest is the spatial distribution of the recorded spectra that are subjected to wavelength and detector-sensitivity calibrations. For diatomic spectra, detector-sensitivity corrections are significant as it is not unexpected to see a variation across several nanometer, or across the vibrational and rotational peaks indicated in the figure. The intensity distribution along the slit height near pixel 512 displays the variation of the rotational spectra. The vertical distributions along the slit are measures for the excitation temperature of CN in the plasma. For time delays within

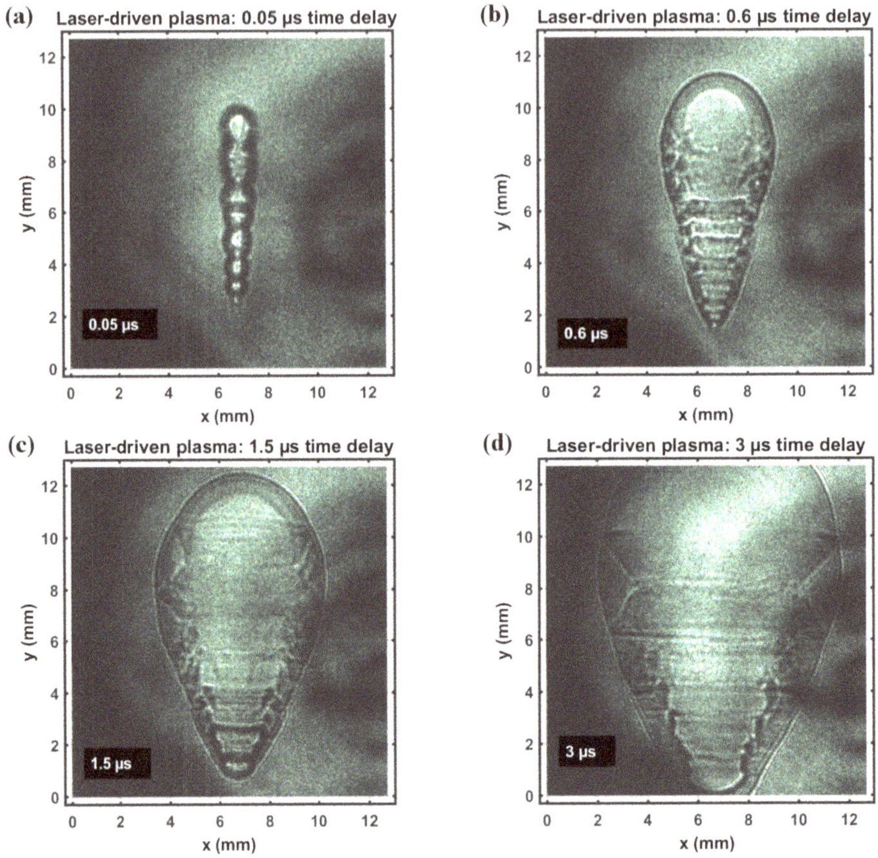

Figure 15.7. Shadow-graphs. Time delays (a) 0.05 μs, (b) 0.6 μs, (c) 1.5 μs, and (d) 3 μs. Multiple breakdown spots lead to stagnation regions. Absorption of the laser beam near the top of the images leads to near spherically- or spheroidally-symmetric expansion. The central vertical line is imaged on the spectrometer slit for capture of spectra along the vertical direction [19].

the first microsecond, the vertical CN spectra indicate expansion dynamics associated with laser-induced optical breakdown. This is further elaborated in the next section.

15.3.4 Abel-inverted CN spectra

Recent experiments [2, 20, 21] explore expansion dynamics of laser-induced optical breakdown in a 1:1 mole ratio CO_2:N_2 mixture at 1 atm. Abel-inversion of the CN data [22] is aimed at obtaining spatial distribution reminiscent of the expanding shockwave. In this section, analysis of the line-of-sight data is elaborated first and then results are presented of the Abel-inverted data.

The experimental arrangement consists of a set of components typical for time-resolved, laser-induced optical emission spectroscopy [9, 20, 23], or nanosecond laser-induced breakdown spectroscopy (LIBS). The experimental series for the

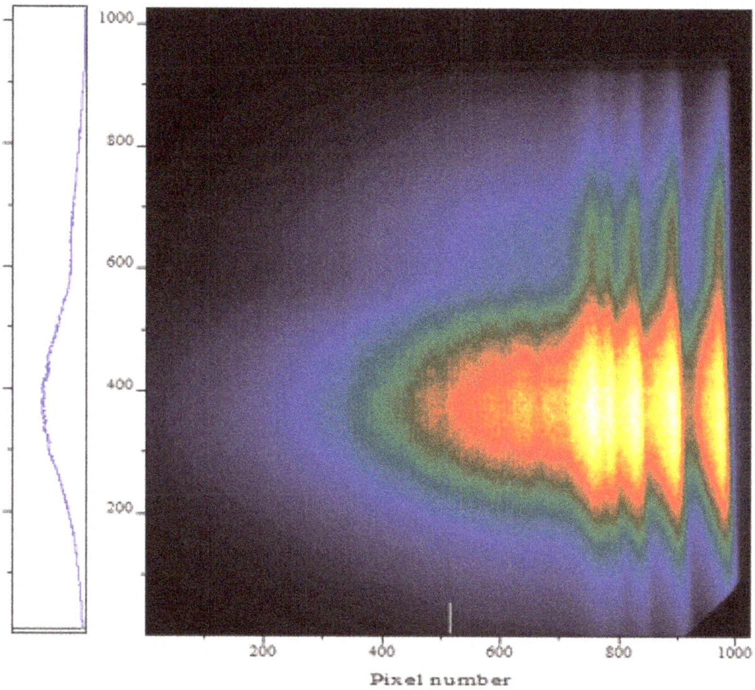

Figure 15.8. Raw CN rotational–vibrational spectra recorded with an ICCD. The horizontal axis corresponds to wavelength, and the vertical axis to the slit height. The four band heads of the CN $\Delta v = 0$ are well developed between pixels 700 and 1000, including undulations between pixels 500 and 700 due to overlapping rotational spectra.

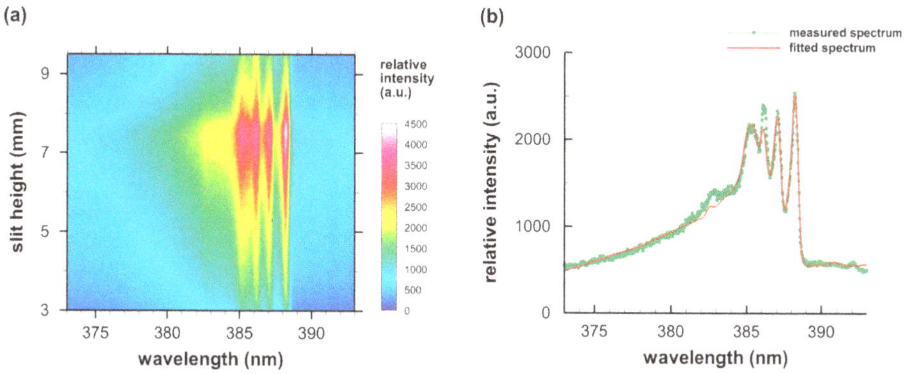

Figure 15.9. Time delay 0.95 µs; (a) pseudo-colored contour map, and (b) measured and fitted spectrum at a slit height of 7 mm, inferred temperature: $T = 9.04$ kK. Reproduced from [21].

separate measurements of CN molecular distribution after optical breakdown includes evacuating the cell to a nominal mercury pump vacuum of 10^{-4} Pa (10^{-6} Torr) and then introducing the N_2:CO_2 mixture. Figure 15.9 illustrates the wavelength-calibrated, sensitivity-corrected, and background-subtracted maps of captured time-resolved data following optical breakdown in the ultra-high pure N_2 and research grade CO_2. And the figure shows a single measured [20] and computed [9] spectrum at the center of the plasma. The fitted spectral resolution, $\Delta\lambda$, amounts to $\Delta\lambda = 0.44$ nm.

Signatures of the 0–0, 1–1, 2–2, 3–3, 4–4 band heads begin to emanate for time delays of the order of 100 ns from optical breakdown. Moreover, the plasma typically propagates toward the top or laser side. The recorded data indicate a ~0.8 mm upward CN-signal propagation in the 370–393.5 nm spectral, ~7 mm object window during the first 5 μs, from a delay of 0.2–5.2 μs.

The detector pixels are binned in four tracks along the slit direction, resulting in obtaining 256 spectra for each time delay. The average of accumulated data from 100 consecutive optical breakdown events is recorded for 21 different time delays using a gate width of 125 ns. In figure 15.9, the vertical axis indicates the slit-height. With 1:2 imaging, and a pixel resolution of 13.6 μm, the clearly discernable plasma size in the cell amounts to ~3 mm. The figure shows that the CN band heads of the $\Delta v = 0$ sequence are well-developed, and it also shows an atomic line near 386.2 nm that likely is the carbon CI 193.09 nm atomic line recorded in second order. The electron density, n_e, can be determined from the Stark full-width at half-maximum, $\Delta\lambda_{\text{Stark}}$, or from the Stark shift, $\delta\lambda_{\text{Stark}}$, of the carbon line [24],

$$\Delta\lambda_{\text{Stark}} = 2wn_e[10^{17}\ \text{cm}^{-3}] \qquad \text{or} \qquad \delta\lambda_{\text{Stark}} = dn_e[10^{17}\ \text{cm}^{-3}], \tag{15.1}$$

where the width parameter, w, and shift parameter, d, amount to $w = 0.00291$ nm and $d = 0.00294$ nm [24], respectively, at an electron temperature of 40 kK. At 200 ns time delay, the full width of ~0.4 ± 0.1 nm in second order reveals an electron density of ~$34 \times 10^{17}\ \text{cm}^{-3}$. The shift of 0.2 ± 0.05 nm in second order yields a consistent result for the electron density. The accuracy of the Stark width parameter is listed at 30% [25].

Figure 15.10 displays measured and fitted spectra at the center and at the top of the spectra at a time delay of 2.2 μs. The plasma shows a spatial temperature variation with the temperature at the center at 8.42 kK and at the top at 9.03 kK. The higher temperature near the edge versus the center indicates higher electron

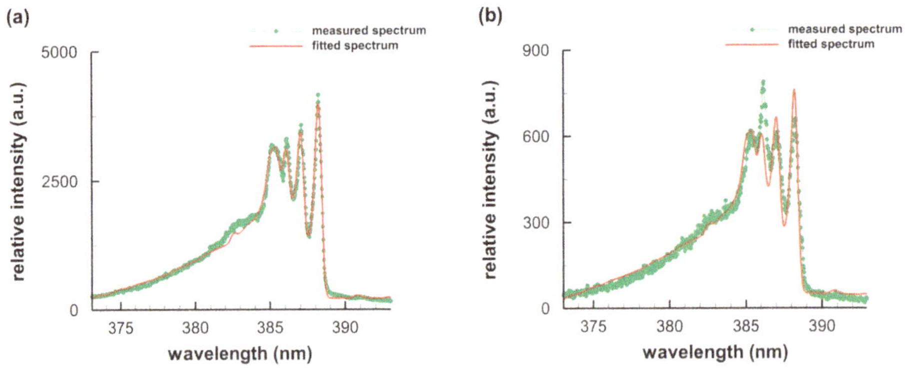

Figure 15.10. Time delay 2.2 μs; (a) center, $T = 8.42$ kK, and (b) top, $T = 9.03$ kK. Reproduced from [21].

density. Additionally, the presence of the CI 193.09 nm atomic line recorded in second order as displayed in figures 15.9(b) and 15.9(b) indicates an electron density that is larger near the edges than at the center of the plasma.

Near the edges of the expanding plasma, the CN signals are lower than near the center. Computations of equilibrium CN distributions versus temperature [20] indicate about a factor of 3 smaller CN mass fractions at 9 kK than at 8 kK. Moreover, application of Abel inverse integral transforms [20] reveals slightly lower CN temperature in the plasma center for time delays of the order of 1 μs. Near the plasma edges at the top and the bottom (see figure 15.9(a)), the signals of the likely atomic carbon line are spectrally wider, show larger shifts, and are stronger than the CN 0–0 band head.

The line-of-sight recorded spectra are further analyzed using Abel inversion techniques. In view of the shadowgraphs for optical air breakdown with 850mJ/pulse 6 ns excitation, spherical symmetry may only be indicated near the top of the images, see figure 15.7. However, as the energy per pulse is reduced by about one order of magnitude, a smaller number of breakdown sites occur. The optical breakdown is initiated near focus of a perfect lens, of course with residual aberration effects, but with diminished forward-cone structure. Hence, the expanding breakdown plasma can be viewed as spherically symmetric. Slight deviation from spherical symmetry, indicated by a forward cone, reflect the amount of plasma generation with above-threshold irradiance.

Figure 15.11 displays Abel-inverted spectra for time delays of 1.2 and 3.2 μs. Clearly, the 1.2 μs spectra indicate diminished signals of CN near the center. The 193.09 nm carbon line, measured in second order, shows a larger width near the edges than at center. This implies larger electron density near the edges than at center, and thus, higher temperature for larger electron density due to isentropic expansion. In other words, hotter CN spectra are measured near the expanding shock wave. In turn, for the 3.2-μs time-delay map, the CN distribution appears flat, but perhaps indicating slightly higher temperature at center than at the edges. There are no signatures from the carbon line for the 3.2-μs data.

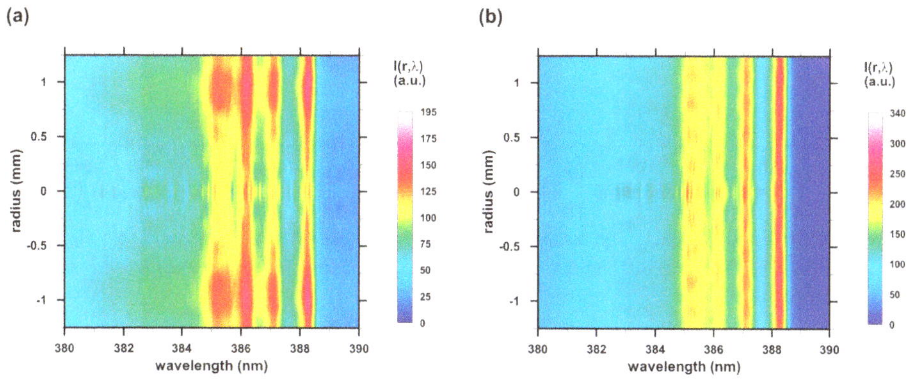

Figure 15.11. Abel-inverted spectra versus radius. Time delays (a) 1.2 μs [2] and (b) 3.2 μs.

Figure 15.12 illustrates measured and fitted CN spectra at the center and at a radius of 0.85 mm for data recorded at 1.2-μs time delay. A modified Boltzmann-plot method [1] is utilized in the recently published program [9] for the determination of the best-fit FWHM, $\Delta\lambda$, and temperature, T, from the entire spectrum. Appendix C further discusses the modified Boltzmann-plot method. The determination of the temperature and density variation of CN and carbon line-width along the slit is a topic of ongoing research. The variation is expected to be analogous to recently reported hydrogen–nitrogen laser-plasma work [26, 27]. Recorded spectra following optical breakdownn in the carbon dioxide–nitrogen mixture at a time delay of 3.7 μs and at 1 mm show lower temperature [20] than that reported here for the 1.2-μs time delay.

The temperature differences in figure 15.12 appear to be rather small, however, a systematic investigation of both the line-of-sight and Abel-inverted CN spectra is expected to reveal similar, relative temperature profiles as obtained in hydrogen–nitrogen laser-plasma experiments [26, 27], i.e. larger density and temperature near the edges than at center, including a temperature drop just inside the shockwave, and minute variations near the center. This trends appears to be indicated as well by the carbon line that is measured in second order.

Figure 15.13 displays the hydrogen-atoms temperature variation along the plasma. The shock wave radius, $R(\tau)$, as a function of time delay, τ, is evaluated with the Taylor–Sedov [19] formula for spherical expansion,

$$R(\tau) = \left(E_p/\rho\tau^2\right)^{1/5}. \tag{15.2}$$

The laser-pulse energy, E_p, for the hydrogen–nitrogen experiments [27] amounts to 0.15 J, and the density of the 1:1 nitrogen hydrogen mixture, ρ, equals 0.37 kg m^{-3}. For $\tau = 3$ μs, one finds for the radius $R = 5.2$ mm.

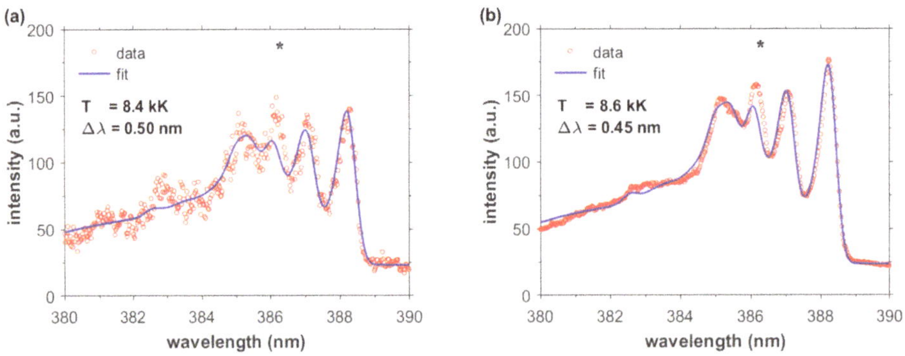

Figure 15.12. Inferred CN excitation temperature (a) at the center, (b) at a radius of 0.85 mm, 1.2-μs time delay, gate 0.125 μs. The asterisks indicate the 386.2 nm position for the 193.1 nm carbon line measured in second order [2].

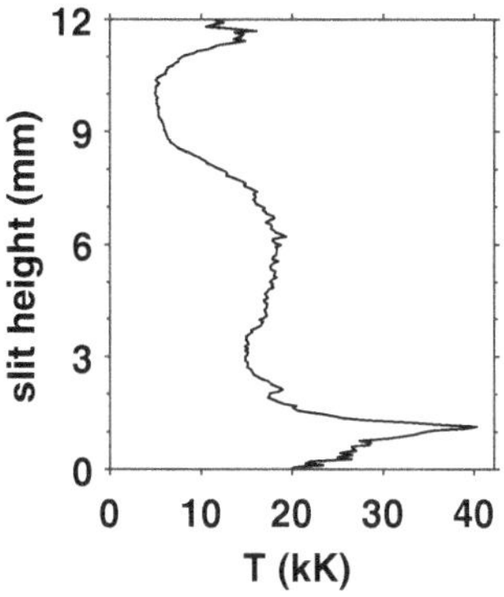

Figure 15.13. Electron temperature from line-of-sight data, 3-μs time delay, 0.025-μs gate. Average temperature: 15.4 kK [27].

References

[1] Hornkohl J O, Parigger C G and Lewis J W L 1991 *J. Quant. Spectrosc. Radiat. Transfer* **46** 405

[2] Parigger C G, Helstern C M and Gautam G 2019 *Atoms* **7** 7030074

[3] Gordon S and McBride B J 1976 *Computer Program for Calculation of Complex Equilibrium Compositions, Rocket Performance, Incident and Reflected Shocks, and Chapman-Jouguet Detonations NASA Lewis Research Center, Interim Revision, NASA Report SP-273* p 273

[4] McBride B J and Gordon S 2005 Computer program on-line 2005 version at https://cearun.grc.nasa.gov/ (last accessed November 26, 2017). *Computer Program for Calculating and Fitting Thermodynamic Functions, NASA RP-1271* p 1271

[5] Parigger C G, Hornkohl J O and Nemes L 2010 *Int. J. Spectrosc.* **2010** 159382

[6] Hornkohl J O and Parigger C G 1996 Boltzmann equilibrium spectrum program (BESP) (accessed January 19, 2016) http://view.utsi.edu/besp/

[7] Nemes L, Keszler A M, Hornkohl J O and Parigger C G 2005 *Appl. Opt.* **44** 3661

[8] Park H S, Nam S H and Park S M 2005 *J. Appl. Phys.* **2005** 113103

[9] Parigger C G, Woods A C, Surmick D M, Gautam G, Witte M J and Hornkohl J O 2015 *Spectrochim. Acta Part B: At. Spectrosc.* **107** 132

[10] Nelder J A and Mead R 1965 *Comput. J.* **7** 308

[11] Salieri F, Quarteroni A and Sacco R 2000 *Numerical Mathematics* (New York: Springer)

[12] Luque J and Crosley D R 1999 *LIFBASE, Database and Sspectral Simulation for Diatomic Molecules (v 1.6)* SRI International Report MP-99-009

[13] Western C M 2017 *J. Quant. Spectrosc. Radiat. Transfer* **186** 221

[14] Western C M 2019 Private Communication

[15] Downs M J and Birch K B 1994 *Metrologia* **31** 315

[16] Parigger C G 2006 Laser-induced breakdown in gases: experiments and simulation *Laser-Induced Breakdown Spectroscopy(LIBS), Fundamentals and Applications* ed A W Miziolek, V Palleschi and I Schechter (New York: Cambridge University Press) ch 4

[17] Kroll N and Watson K M 1972 *Phys. Rev.* A **5** 1883

[18] Thiyagarajan M and Thompson S 2012 *J. Appl. Phys.* **111** 073302

[19] Gautam G, Helstern C M, Drake K A and Parigger C G 2016 *Int. Rev. At. Mol. Phys.* **7** 45

[20] Parigger C G, Helstern C M, Drake K A and Gautam G 2017 *Int. Rev. At. Mol. Phys.* **8** 53

[21] Helstern C M and Parigger C G 2019 *J. Phys.: Conf. Ser.* **1289** 012016
[22] Parigger C G, Surmick D M, Helstern C M, Gautam G and Bol'shakov A A 2019 Molecular laser-induced breakdown spectroscopy *Laser-Induced Breakdown Spectroscopy* ed J P Singh and S N Thakur (New York: Elsevier) ch 7
[23] Parigger C G, Woods A C, Witte M J, Swafford L D and Surmick D M 2014 *J. Vis. Exp.* **84** 51250
[24] Griem H 1974 *Spectral Line Broadening by Plasma* (New York: Academic)
[25] Konjević N, Lesage A A, Fuhr J R and Wiese W L 2002 *J. Phys. Chem. Ref. Data* **31** 819
[26] Parigger C G 2019 *Atoms* **7** 7030061
[27] Parigger C G 2019 *Contr. Astron. Obs. Skalnaté Pleso* in press

IOP Publishing

Quantum Mechanics of the Diatomic Molecule with Applications

Christian G Parigger and James O Hornkohl

Chapter 16

Diatomic carbon, C_2

The prediction of diatomic molecular carbon spectra has been a significant goal [1] in the research efforts that are discussed in this book. Plasma emission spectra serve as a gauge for the accuracy of the computed spectra. This chapter focuses on aspects of diatomic carbon molecular spectroscopy that were communicated at two recent line shape conferences [2, 3].

In the laboratory, laser-induced breakdown spectroscopy (LIBS) is a technique utilized for analyzing a substance's composition [4]. It is a non-invasive and non-contact technique that only requires a small amount of the substance to be useful while still allowing for accurate determination of unknown material [5]. This becomes especially useful when the substance itself or its environment is hazardous in nature. However, in order to be of use as a diagnostic tool, well-tested models are required in conjunction with interpretation of well-developed molecular C_2 spectra such as recombination spectra recorded following laser-induced optical breakdown [6]. Recent investigations [2, 7–9] carbon Swan spectra emissions from laser-induced plasma.

16.1 Analysis of C_2 in Nd:YAG laser-plasma

Carbon Swan spectra are observed following laser ablation of graphene in laboratory air [2] or on solid targets and in gas-phase optical breakdown [10, 11]. Previous experiments showed temperatures that ranged from 4.5 to 7.5 kK for the $\Delta v = 0$ transition and 4.2 to 4.5 kK for the $\Delta v = -1$ transition for time delays of the order of 1.6 μs to 70 μs. This experiment explored in greater detail time delays >10 μs for both molecular bands. Temperatures were found to be similar, ranging from 4.5 to 6.7 kK for the $\Delta v = 1$ transition and 3.2 to 5.5 kK for the $\Delta v = -1$ transition. Figure 16.1 illustrates C_2, $\Delta v = -1$ emission spectra recorded at a delay of 30 μs in CO gas at 1 atm.

doi:10.1088/978-0-7503-1989-8ch16

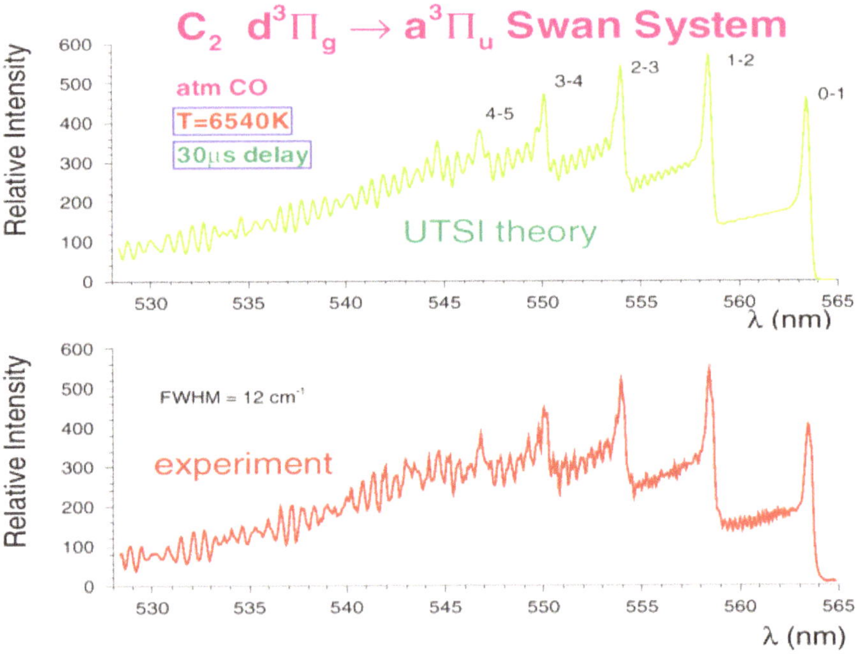

Figure 16.1. Measured and computes C_2, $\Delta v = -1$ squence. Time delay 30 μs, temperature 6.54 kK, spectral resolution 0.36 nm (12 cm^{-1}). The spectrum is measured following optical breakdown in CO gas at 1 atm. Reprinted from [6], with permission of Elsevier.

16.2 Detailed fitting of C_2 spectra

Efforts are communicated to investigate the applicability of the local thermodynamic equilibrium (LTE) assumption. Comparisons are discussed in view of previous work that utilized Stark broadening of the H_β line, confirming LTE for delays <10 μs, yet further research is needed for longer time delays. Recent experiment utilizes a Nd:YAG pulsed laser operating at 1064 nm with 13 ns, 190 mJ/pulse to generate laser-induced breakdown. The emitted spectra were dispersed with an 1800 grooves/mm grating in a HR640 Jobin-Yvon 0.64-m spectrometer and then recorded with an Andor iSstar ICCD. Measurements were time-resolved with varying time delays (10–100 μs) and corresponding gate widths (1–20 μs). Spectra were sensitivity and wavelength calibrated with a standard tungsten and mercury lamp, respectively.

For diatomic molecules, spectral predictions require a temperature along with a set of line-strengths [1], for the allowed spectral transitions. In order to fit the spectra, the Nelder–Mead algorithm is utilized [12] through the so-called Nelder–Mead temperature (NMT) program [13]. The NMT program generates a single temperature spectral fit for the recorded data. This fit, within the accuracy and precision bounds, is also used to refine the calibration of the data. This is repeated several times, with each run increasing from a linear, quadratic, to finally a cubic

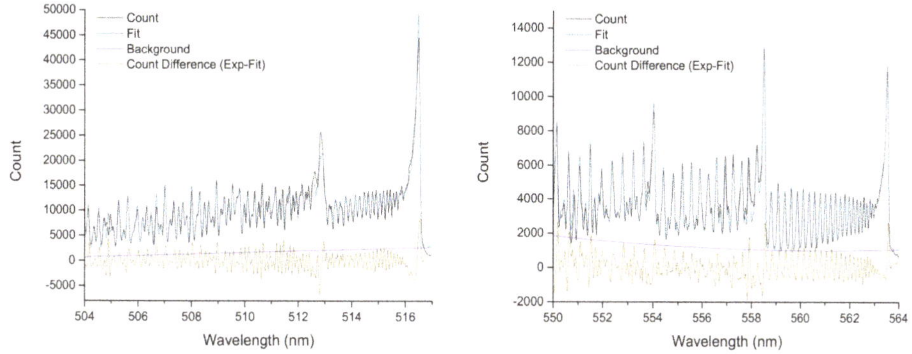

Figure 16.2. Measured and fitted C_2 spectra: 40 μs delay, 5 μs gate width. Temperatures $T = 5.5$ kK and $T = 4.4$ kK, left and right. Reproduced from [2]. © IOP Publishing Ltd. CC BY 3.0.

calibration. After that, the program is run three times in succession to ensure accuracy of the computed temperature.

Figures 16.2 and 16.3 are representative of the results from the NMT program for carbon Swan spectra of the $\Delta v = 0$ and $\Delta v = -1$ bands. Seen in both are the clear vibrational peaks and underlying rotational structure. No atomic lines are present. Note, there is good matching between the experimental and theory spectra as the count difference between both is generally within the computed background line. However, the discrepancy in temperature is in part due to the different backgrounds that were used in the analysis of the $\Delta v = 0$ and $\Delta v = -1$ analysis. Moreover, the data were collected in separate experimental runs.

Figure 16.4 shows a 2D image of the C_2 spectra for the $\Delta v = 0$ band. Calculated temperatures range from 6600 to 7700 K, increasing toward 7 mm, with a dip just under it at 6 mm. Also of note is the slight vertical symmetry.

For the theory models, a single temperature is used for fitting with assumption of LTE. For early time delays of the order of 1 μs with laser energy per pulse of the order of 100 mJ, time-delayed spectra indicate LTE due to the presence of H_β [8]. Past that, there is some indication for LTE at later delays: the experimental spectra

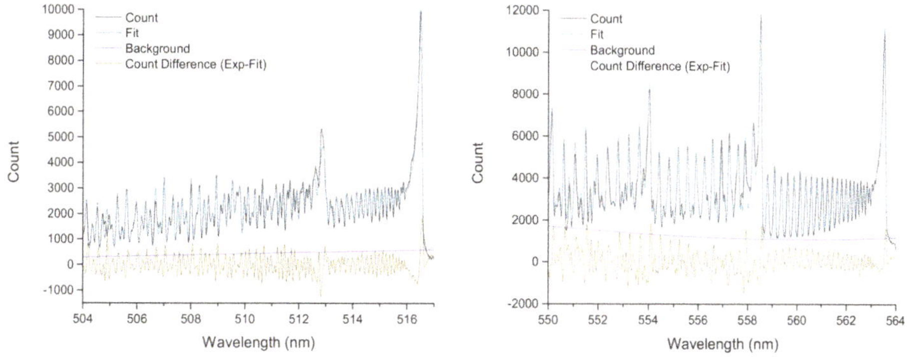

Figure 16.3. 70 μs delay, 10 μs gate width. $T = 5.8$ and 3.2 kK, left and right. Reproduced from [2]. © IOP Publishing Ltd. CC BY 3.0.

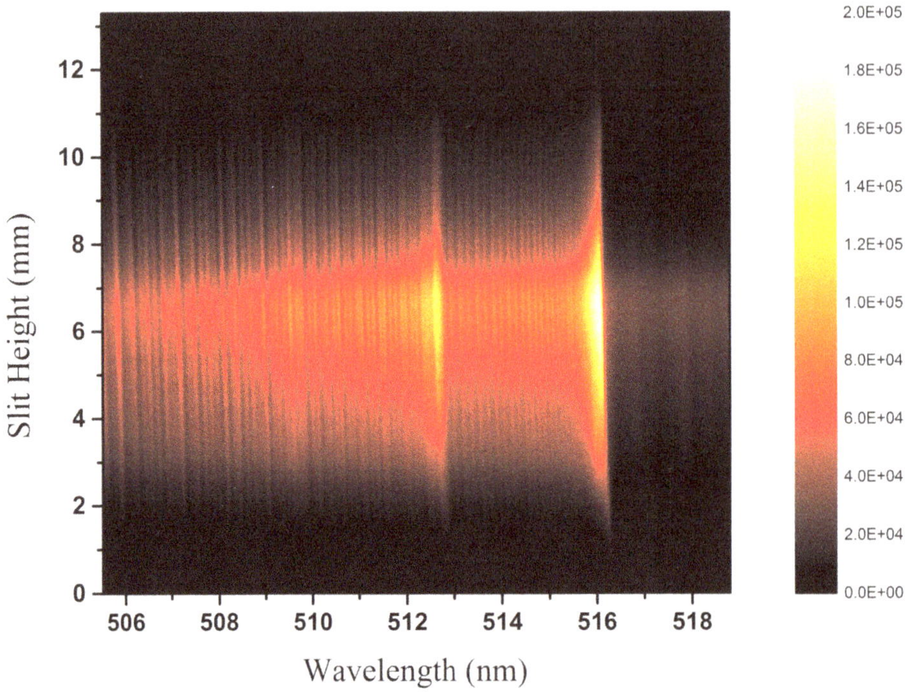

Figure 16.4. Raw image of two dimensional C_2 $\Delta v = 0$ spectra at 15 μs delay and 5 μs gate width. The FWHM averaged to 0.13 nm for each of the individual spectral lines. Reproduced from [2]. © IOP Publishing Ltd. CC BY 3.0.

show good agreement with computed theory spectra within background variations. However, there is no definite indication of LTE at later delays. As the electron density diminishes, H_β emissions disappear in the measured spectra past 10 μs. There is some indication that LTE is absent since there appear to be different temperature ranges for the different wavelength bands, as seen in comparing the $\Delta v = 0$ and $\Delta v = -1$ sequences that were collected from separate graphene laser ablation events.

From the recorded 2D data, the plasma shows a higher temperature toward the top of the recorded emission spectra near 7 mm but is cooler away from it. Just below that region, there appear to be significant variations in inferred temperature. These temperature differences could indicate non-LTE conditions. Consequently, exploration and utilization of Abel- and Radon-techniques [14] in better analyzing the plasma are called for.

16.3 Superposition spectra of hydrogen and carbon

Superposition spectra of hydrogen and carbon Swan bands [3] occur in hydrogen–carbon plasma generated by high intensity laser radiation. Investigated are the H_α, H_β, and also H_γ lines and their characteristics as well as broadening and shifts. This article elaborates on previously communicated emission spectroscopy data of the hydrogen Balmer series [15–19]. Laser-induced breakdown spectroscopy (LIBS) historically embraces elemental analysis, or atomic spectroscopy, and to a lesser extent molecular spectroscopy. Time-resolved diagnostics allow us to record atomic emission profiles separate from molecular recombination and/or excitation; however, atomic and molecular species can occur simultaneously within a selected spectral window, for our expanding methane experiments typically on the order of 1–2 μs after optical breakdown.

Following the invention of laser devices, laser-induced breakdown spectroscopy (LIBS) evolved into one of the favorite topics not only in atomic and chemical physics but also in chemistry, biology, environmental science, material engineering, and recently space exploration. Modeling of the hydrogen line shapes for the diagnostic for International Thermonuclear Experimental Reactor (ITER) is an important ongoing investigation, and it is based on the Balmer lines of the hydrogen isotopes. The use of hydrogen Balmer lines has been employed to measure the temperature and electron density in plasmas, including several theoretical approximations [20–22]. In typical LIBS experiments, the main broadening mechanism of spectral lines is the Stark effect. Line broadening has been studied comprehensively by Griem [23]. A large, linear Stark effect in hydrogen makes its lines attractive for recent research [24]. According to theory, the shapes and shifts of plasma-broadened isolated lines are mainly determined by electron impact with the radiating atom or ion, and a smaller contribution arises from the electric micro-fields generated by essentially static plasma ions. In addition, at a given electron number density of the plasma, the hydrogen lines are broader with respect to the lines of other elements. Doppler broadening has information about velocity distribution, and Stark broadening gives information on electron density [25]. These features make their use particularly interesting in LIBS experiments performed with typically 0.1 nm spectral resolution spectrometers where the instrumental broadening introduced by the spectrometer might be comparable, or even larger, than the broadening of the emission lines of elements different from hydrogen. Moreover, the Stark broadening of the emission lines is independent of the fulfillment of local thermodynamic equilibrium (LTE) conditions [26], and this aspect makes this approach for the electron density measurement an interesting and powerful tool. Measurements of line shapes are important in plasma diagnostics, allowing us to investigate fundamental atomic structure as well as inferring plasma temperature and electron density. In particular, the H_β line is one of the most studied lines, and its use has become a standard technique in plasma spectroscopy [27].

In previous studies by Parigger *et al* [28], analysis of Stark broadening and shifts in measured H_α and H_β spectra, combined with Boltzmann plots from H_α, H_β, and H_γ lines to infer the temperature, have been discussed for the electron densities in the range of 10^{16}–$10^{19}\mathrm{cm}^{-3}$, and for the temperatures in the range of 6–100 kK. Lines H_α, H_β, and H_γ were recorded at different delay times from optical breakdown. The electron density was evaluated using different theoretical approaches that are commonly employed for plasma diagnostics [17]. We previously reported measurements [15–19] and comprehensive studies of the diagnostic application of Balmer lines.

For generation of a micro-plasma, a laboratory Nd:YAG laser operated at the fundamental wavelength of 1064 nm was used. For the hydrogen plasma investigations, a Continuum YG680S-10 Nd:YAG with 150 mJ energy per pulse and 7.5 ns pulse duration was focused to typically 1.4 TW cm^{-2} in a pressure cell that was filled with gaseous hydrogen to a pressure of 810 ± 25 Torr (1.07 × 10^5 Pa) and 1010 ± 25 Torr (1.33 × 10^5 Pa), subsequent to evacuation of the cell with a diffusion pump. Spectrometer and detector arrangements are discussed in the Parigger and Oks review article [15]. Recording of shadow graphs (not illustrated here) was accomplished with Coherent Infinity 40-100 Nd:YAG laser with 50 mJ and 300 mJ pulse energy and 3.5 ns pulse width was focused in laboratory air to an irradiance of typically 10 TW cm^{-2}. Perpendicular to the laser beam generating LIB, we typically employed pulsed laser radiation from a 308 nm XeCl excimer laser.

In the expanding methane experiments [18], we used a 1/2 m model 500 SpectraPro Acton Research Corporation spectrometer together with an intensified linear diode array (model 1460 Princeton Applied Research detector/controller optical multichannel analyzer). Averaged over 100 individual LIB events, the captured time-resolved data (measurement window of 0.1 μs) from the methane breakdown events were detector-noise/background corrected and wavelength and detector sensitivity calibrated using standard calibration lamps.

Measurements of individual profiles of the Balmer series lines H_α, H_β, and H_γ allows us to infer electron excitation temperature. However, the inferred temperature will show error-bar contributions due to the broadened atomic emissions. Presence of the molecular emissions will add to the background for the hydrogen Balmer series diagnostic. We treat the H_β and H_γ lines as background contribution to extract molecular temperature information for analysis of molecular excitations [19]. Time resolved spectroscopy measurements of hydrogen-alpha, -beta, and -gamma emissions were performed in an expanding flow of methane gas. Hydrogen Stark widths were used to characterize the plasma decay during the first 2 μs after laser-induced optical breakdown [18]. Analysis of the recorded hydrogen spectra include measurements of full-width-half-maximum (FWHM) and full-width-half-area (FWHA) to infer electron density N_e [16]. Temperature was determined from incomplete hydrogen Balmer Series lines [17]. Yet for time delays on the order of 1 μs, molecular emissions can be recognized from the C_2 Swan system, superposed to the H_β and H_γ Balmer series lines. C_2 emissions are clearly recognizable for time delays of the order of 2 μs [3]. Figures 16.5 and 16.6 show these results.

Figure 16.6 indicates the presence of C_2 molecular emissions for a delay time of $\Delta\tau = 2.1$ μs. Temperature inferred from the fitting of the C_2 'fingerprints' revealed molecular temperatures in the range of 4–6 kK [19]. The H_β profile at a delay time of $\Delta\tau = 2.1$ μs shows an electron density of 0.5 × 10^{17} cm^{-3}. The electron excitation temperature for delay times on the order of 1.1–2.1 μs amounts to 13–11 kK [17], while the molecular temperature of the residual carbon, see figure 13.2, is determined to be 5.2 kK.

The H_β profiles typically show a double-peak structure consistent with computed line profiles that appear asymmetric early in the plasma decay. The major broadening mechanism here is due to the Stark effect, which should cause a Lorentzian

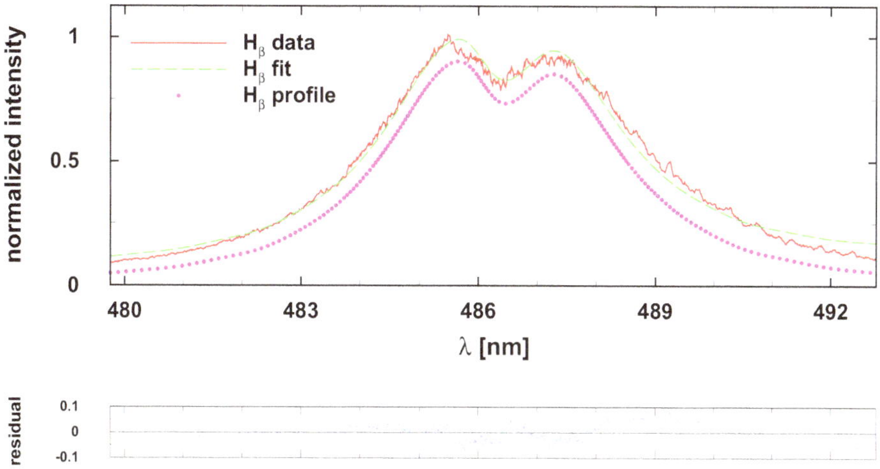

Figure 16.5. H_β recorded and fitted spectra. Delay time $\Delta\tau = 1.1$ μs. $N_e = 1.0 \times 10^{17}$ cm^{-3}. Reproduced from [3], © IOP Publishing Ltd. CC BY 3.0.

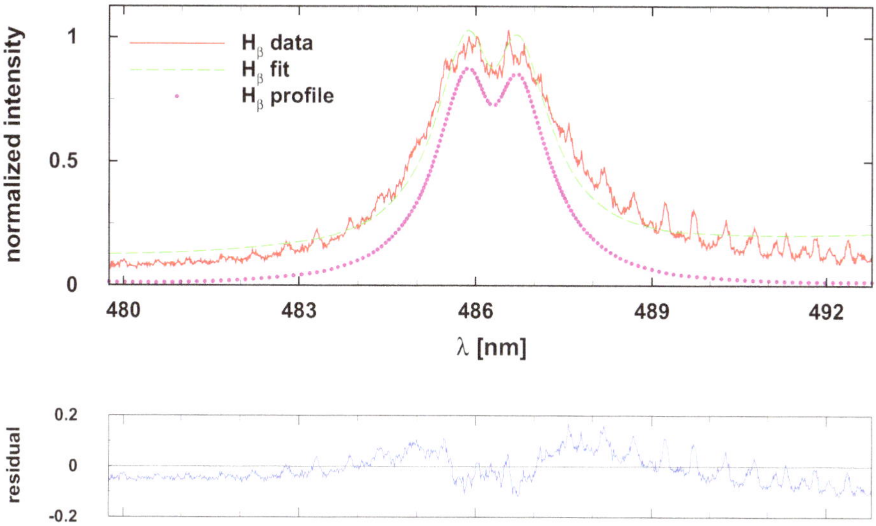

Figure 16.6. H_β recorded and fitted spectra. Delay time $\Delta\tau = 2.1$ μs. $N_e = 0.5 \times 10^{17}$ cm^{-3}. Reproduced from [3], © IOP Publishing Ltd. CC BY 3.0.

profile, and the dip at the line center is due to absence of computed H_β shifts [29, 30] and in part due to electrostatic interactions with ions [31]. The theory for widths of the hydrogenic lines is mostly accurate and best agrees for H_β, where the overall error is about 5%. For hydrogenic systems, a relatively large portion of the broadening is due to quasi-static effects.

Electron density in plasma can be found by measuring Stark broadening, and this is the most precise technique for determining the electron density N_e in plasma. This broadening will become significant for $N_e > 10^{15}\,\text{cm}^{-3}$. Stark broadening is due to collision between radiating species with charged species, like electrons and ions. Not only the H_β Stark width, but also the separation of the H_β double-peak structure is a measure for the free electron density [28]. Oks' theory of Stark widths incorporates several major theoretical advances and new phenomena in the Stark broadening of hydrogen lines in plasmas [32]. The Stark linewidth $\Delta\lambda_{\text{FWHM}}$ at full-width-half maximum (FWHM) can be extracted from the measured line width $\Delta\lambda_{\text{observed}}$ by subtracting instrumental $\Delta\lambda_{\text{instrumental}}$ and Doppler line broadening $\Delta\lambda_{\text{Doppler}}$ [33]. A large electron density ($N_e > 10^{17}\,\text{cm}^{-3}$) causes Stark broadening that is typically much larger than employed instrumental resolution and Doppler broadening. Experimental studies of aluminum laser ablation [34] within an evacuated cell reveals the limitation of Balmer series diagnostic for electron density and temperature. Early in the plasma decay, the H_α line becomes discernable at delays on the order of 0.025–0.050 μs after LIB, showing an electron density of $10^{18}\,\text{cm}^{-3}$. The H_β line appears recognizable at delays on the order of 0.075–0.100 μs, showing an electron density of $4 \times 10^{17}\,\text{cm}^{-3}$.

A plasma's temperature and density are important for its stability. For high-temperature nuclear fusion plasma, non-invasive plasma diagnostics (like optical emission spectroscopy) are necessary [35]. A typical method to infer plasma temperature is to construct a so-called Boltzmann plot [36] including consideration of non-uniformity of laser-induced plasma [37]. Relative ion populations in local thermodynamic equilibrium (LTE) can be computed using the Saha (or Saha–Eggert) equation. However, when applying the Saha equations to obtain temperature, called the temperature of the ionization equilibrium, relative intensities of lines from different ion stages of the same atom (occasionally different atoms) need to be measured [38].

The recorded emission spectra of the hydrogen Balmer series allow us to infer electron number density, using previously tabulated values. The displayed hydrogen emission profiles are also convolved with the Doppler width that amounts to 0.05 nm at 10 kK and 0.15 nm at 100 kK. There is a substantial red shift of the H_α line profile from its unperturbed position for N_e of the order of $10^{18}\,\text{cm}^{-3}$. Larger red shifts in the forward region of the breakdown plasma correspond to higher temperature and also electron density, therefore, we infer that there is an electron (or excitation) temperature gradient across the length of the plasma. Temperatures on the order of 100 kK can be found early in the plasma decay [28]. Use of Oks' theory allowed us to enhance significantly the agreement of theory and our experimental results.

Superposition spectra occur due to recombination or due to onset of chemical reactions. Analysis of both atomic and molecular emission spectra following laser-induced optical breakdown utilize molecular diagnostics based on line-strength files. The illustrated spectroscopic data display both the Balmer series hydrogen beta line and molecular C_2 Swan band presence. The molecular excitation temperature is determined using modified Boltzmann plots and fitting of spectra from selected molecular transitions [1].

References

[1] Hornkohl J O, Nemes L and Parigger C G 2009 Spectroscopy, dynamics and molecular theory of carbon plasmas and vapors *Advances in the Understanding of the Most Complex High-Temperature Elemental System* ed L Nemes and S Irle (Singapore: World Scientific) ch 4

[2] Witte M J and Parigger C G 2014 *J. Phys.: Conf. Ser.* **548** 012052

[3] Parigger C G, Woods A C and Rezaee M R 2012 *J. Phys.: Conf. Ser.* **397** 012022

[4] Parigger C G 2013 *Spectrochim. Acta Part B: At. Spectrosc.* **78–79** 4

[5] De F C, Lucia, Harmon R S, McNesby K L, Winkel R J and Miziolek A W 2003 *Appl. Opt.* **42** 6148

[6] Parigger C G, Plemmons D H, Hornkohl J O and Lewis J W L 1994 *J. Quant. Spectrosc. Radiat. Transfer* **52** 707

[7] Witte M J, Parigger C G, Bullock N A, Merten J A and Allen S D 2014 *Appl. Spectrosc.* **68** 367

[8] Witte M J and Parigger C G 2013 *Int. Rev. At. Mol. Phys.* **4** 63

[9] Parigger C G, Woods A C, Witte M J, Swafford L D and Surmick D M 2014 *J. Vis. Exp.* **84** 51250

[10] Parigger C G, Hornkohl J O, Keszler A M and Nemes L 2003 *Appl. Opt.* **42** 6192

[11] Nemes L, Keszler A M, Hornkohl J O and Parigger C G 2005 *Appl. Opt.* **44** 3661

[12] Woods A C, Parigger C G and Hornkohl J O 2012 *Opt. Lett.* **37** 5139

[13] Parigger C G, Woods A C, Surmick D M, Gautam G, Witte M J and Hornkohl J O 2015 *Spectrochim. Acta Part B: At. Spectrosc.* **107** 132

[14] Gornushkin I B, Merk S, Demidov A, Panne U, Shabanov S V, Smith B W and Omenetto N 2012 *Spectrochim. Acta Part B: At. Spectrosc.* **76** 203

[15] Parigger C G and Oks E 2010 *Int. Rev. At. Mol. Phys.* **1** 13

[16] Parigger C G 2010 *Int. Rev. At. Mol. Phys.* **1** 129

[17] Parigger C G, Woods A C and Hornkohl J O 2011 *Int. Rev. At. Mol. Phys.* **2** 77

[18] Parigger C G, Dackman M and Hornkohl J O 2008 *Appl. Opt.* **47** G1

[19] Parigger C G, Woods A C and Hornkohl J O 2012 *Appl. Opt.* **51** B1

[20] Aragón C and Aguilera J A 2010 *Spectrochim. Acta B: At. Spectrosc.* **65** 395

[21] EL Sherbini A M, Hegazy H and EL Sherbini T M 2006 *Spectrochim. Acta B: At. Spectrosc.* **61** 532

[22] Detalle V, Héon R and Sabsabi M 2001 *Spectrochim. Acta B: At. Spectrosc.* **56** 1011

[23] Griem H 1974 *Spectral Line Broadening by Plasma* (New York: Academic)

[24] Konjević N 1999 *Phys. Rep.* **316** 339

[25] Chen F F 1995 *Introduction to Plasma Physics* (New York: Springer)

[26] Cristoforetti G, De Giacomo A, Dell'Aglio M, Legnaioli S, Tognoni E, Palleschi V and Omenetto N 2010 *Spectrochim. Acta B: At. Spectrosc.* **65** 86

[27] Djurović S, Ćirišan S, Demura A V, Demchenko G V, Nikolić D, Gigosos M A and González M Á 2009 *Phys. Rev.* E **79** 046402

[28] Oks E, Parigger C G and Plemmons D H 2003 *Appl. Opt.* **42** 5992

[29] Schrödinger E 1926 *Ann. Phys.* **385** 437

[30] Schrödinger E 1926 *Phys. Rev.* **28** 1049

[31] Laux C 2003 *Plasma Sources Sci. Technol.* **12** 125

[32] Oks E 2006 *Stark Broadening of Hydrogen and Hydrogenlike Spectral Lines on Plasmas: The Physical Insight* (Oxford: Alpha Science)

[33] Camacho J J, Díaz L, Santos M and Poyato J M L 2011 *Spectrochim. Acta Part B: At. Spectrosc.* **66** 57
[34] Parigger C G, Hornkohl J O and Nemes L 2007 *Appl. Opt.* **46** 4026
[35] Kim J-H and Lee H-J 2006 *J. Korean Phys. Soc.* **49** S184
[36] Aydin Ü, Roth P, Gehlen C D and Noll R 2008 *Spectrochim. Acta Part B: At. Spectrosc.* **63** 1060
[37] Gornushkin I B, Shabanov S V, Merk S, Tognoni E and Panne U 2010 *J. Anal. Atom. Spectrom.* **25** 1643
[38] Cremers D E and Radziemski L J 2006 *Handbook of Laser-Induced Breakdown Spectroscopy* (New York: Wiley)

Chapter 17

Aluminium monoxide, AlO

This chapter on aluminium monoxide elaborates aspects for two detailed applications, namely, measurement and accurate analysis of laser ablation from alumina [1], Al_2O_3, and inferences from emission spectroscopy in aluminium combustion. The former experiments were conducted in the CLA at UTSI, and the latter were conducted in collaborative research at Sandia National Laboratories, Albuquerque, NM. This chapter also discusses work communicated at a recent spectral line shape conference [2].

In-depth studies of AlO spectra from alumina ablation included measurements and analysis with a modified diatomic Boltzmann plot and with the Nelder–Mead fitting algorithm of the entire spectrum [1]. Figure 17.1 displays a typical AlO overview spectrum and the corresponding fitted spectrum.

Measurement of the AlO sequences and progression [3] is accomplished using 266 nm radiation from a frequency quadrupled Nd:YAG laser. At the alumina target, an irradiance of 5 GW cm^{-2} was available at the target surface. The resulting plasma from the laser-surface interaction was imaged onto the slit of a 0.275-m Jarrell-Ash spectrometer. A 350 nm cut-on filter was used to effectively reduce second-order interferences. The AlO spectrum was measured by the use of an intensified 1024-pixel linear diode array (LDA) and an optical multichannel analyzer (OMA) that was synchronized to 50 Hz. The linear diode array was read synchronous to the laser repetition rate, and four additional readouts were used to reduce dark-count buildup during the laser's off cycle [1]. Extensive analysis, including conditioning the data with a 15 point 6 degree Savitzky–Golay filter, revealed standard deviations of the order of 1 per cent for the inferred temperature. Moreover, the modified diatomic Boltzmann plot (DBP) and the Nelder–Mead temperature (NMT) fitting routines agreed within 3 per cent. The DBP utilizes prominent peaks rather than fitting the entire AlO spectrum, hence DBP may be better suited than NMT.

doi:10.1088/978-0-7503-1989-8ch17

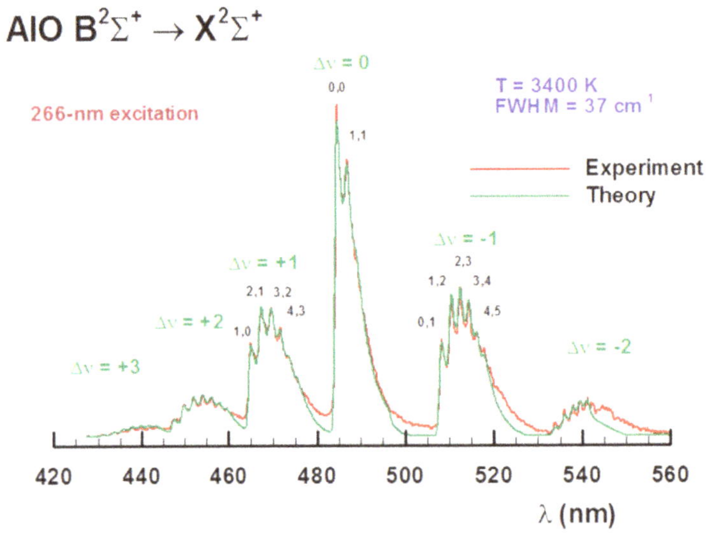

Figure 17.1. Experimental and fitted AlO spectra [1] at a spectral resolution of 0.89 nm ($37\ \text{cm}^{-1}$), 20-μs time delay after plasma generation, 10-μs gate width.

In recent work [2], the aim is characterization of the temperature decay of laser-induced plasma near the surface of an aluminium target from laser-induced breakdown spectroscopy measurements of aluminium alloy sample. Laser-induced plasma are initiated by tightly focussing 1064 nm, nanosecond pulsed Nd:YAG laser radiation. Temperatures are inferred from aluminium monoxide spectra viewed at systematically varied time delays by comparing experimental spectra to theoretical calculations with a Nelder–Mead algorithm. The temperatures are found to decay from 5173 ± 270 to 3862 ± 46 K from 10 to 100 μs time delays following optical breakdown. The temperature profile along the plasma height is also inferred from spatially resolved spectral measurements and the electron density is inferred from Stark broadened H_β spectra.

17.1 Laser-induced breakdown spectroscopy

Time-resolved laser-induced breakdown spectroscopy (LIBS) is a valuable experimental tool for determining key plasma parameters, such as the temperature and electron number density. The laser-induced plasma created following breakdown has multiple applications, including use in forensics, geophysics, and, of particular interest, combustion diagnostics. For the purposes of studying combustion, aluminium LIBS studies are of significant as aluminium is commonly found in solid rocket propellants and high explosives, where the goal is commonly to infer temperature [4, 5]. Determinations of the electron number density of aluminium laser-induced plasma also proven to be useful in diagnostic applications, such as with aluminum alloys [6, 7]. In previous time-resolved, aluminium laser ablation studies, the temperature of laser-induced plasma was determined from Boltzmann plots of aluminium atomic emissions and were found to be in the 4000–4250 K range [4, 8].

The plasma temperature may also be determined from the temperature dependence of diatomic molecular spectra such as aluminium monoxide (AlO). Temperature results from previous studies were found to be 4000 K at 50 μs [9] and 4880 K at 45 μs [10] from AlO emissions. Electron densities were also inferred in reference [9] and were found to be 1023 m^{-3}. In this work, we seek to use AlO diatomic molecular emissions to characterize the temperature decay of laser-induced plasma near the surface of an aluminium target in laboratory air over a broad range of times following optical breakdown using time-resolved LIBS measurements. Inferences are also made on the temperature distribution along the plasma height and the electron number density.

17.2 Experimental details for AlO measurements

Time-resolved spectral measurements are used to characterize the temperature decay of laser-induced plasma initiated near the surface of an aluminium target. Laser-induced plasma are formed by tightly focussing laser radiation from a Q-switched, Nd:YAG laser operating in its fundamental mode with a 12 ns pulse width and an energy per pulse of 190 mJ. Line of sight spectral measurements are made with a Jobin-Yvon HR640 spectrometer with 0.1 nm spectral resolution coupled to gated, intensified detectors. An intensified, linear diode array coupled to an optical multichannel analyzer (Princeton Instruments 1460) was used to collect AlO spectra from 10 to 100 μs time delays in 5 μs steps with gate widths of 5, 10 and 20 μs for 10–20, 20–100, and 70–100 μs delays, respectively. A gated ICCD was used to infer the temperature dependence along the plasma height at a time delay of 60 μs with a 6 μs gate width. Time resolution was achieved by synchronizing the laser repetition rate with the measurement rate of the gated detectors with the use of waveform, pulse, and delay generators. Reference [8] provides a detailed description of the apparatus [11]. Prior to analysis, all spectra were properly calibrated for detector response and background sensitivity.

17.3 Selected results

To infer the temperature as a function of time following optical breakdown, spectra collected with intensified, linear diode array were fit to theoretical calculations of the AlO $B^2\Sigma^+ \rightarrow X^2\Sigma^+$ $\Delta v = 0$ transition. Comparisons were made with a Nelder–Mead fitting algorithm, which was utilized for its ability to fit multiple parameters simultaneously, including a variable baseline offset. The Nelder–Mead algorithm is a minimization method in which a geometric simplex is reduced in size until a satisfactory tolerance level is reached [12, 13]. The theoretical spectra are calculated from tables of accurately calculated line strengths, where the line strength is calculated in its factorized form from the electronic transition strength, the Hönl–London factor, and Franck–Condon factor [1, 14, 15]. The baseline offset of all fitted AlO spectra were quadratic in nature and errors in the inferred temperatures were found by considering variations of the spectral resolution and baseline offset. Analysis showed that the error in the inferred temperatures was in the 45–75 K range, with the exception of the inference made at 10 μs with an error of 236 K. This

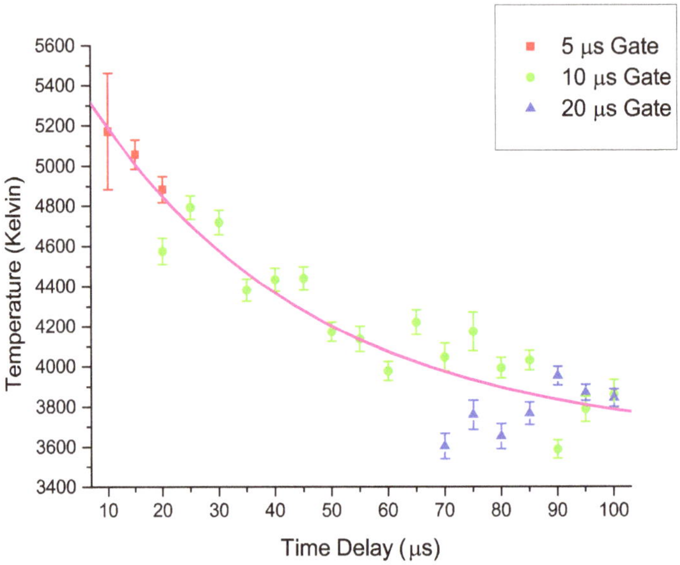

Figure 17.2. The inferred temperatures as a function of time with an exponential decay trend line. Reproduced from [2], © IOP Publishing Ltd. CC BY 3.0.

resulted from the presence of H_β superposition spectra. The results of the temperature analysis are shown in Figure 17.2 with an exponential decay trend. The trend line shows that the plasma temperature decay slows at later times as was expected. To infer the temperature dependence along the plasma height, spatially resolved spectra with 0.013 mm resolution were fit using the Nelder–Mead algorithm. The temperature at the center of the plasma was found to be 5780 ± 236 K at 5.2 mm slit height and decreased to 5396 ± 315 K at 6.5 mm. The temperature was found to increase above the plasma edge to a value of 5820 ± 230 K at a slit height of 9.1 mm. The larger temperatures and errors are likely result from non-optimal plasma and detector conditions; however, the temperature profile suggests that aluminium particles are combusting in a plume of above the plasma, given the ignition temperature of aluminium is 2750 K and the inferred temperatures from the linear diode experiment.

At time delays of 10, 15, and 20 μs following breakdown, superposition of Balmer series H_β spectra with AlO measurements were observed. The H_β spectra may be extracted and used to infer the electron number density from the Stark broadened emission. The line width of the spectra is taken to be due to Stark effects and is inferred by fitting a Lorentzian line profile to the H_β spectra. The Stark widths are used with the empirical theory of Konjević [16] and the convergent theory of Oks [17] to infer the electron number density, N_e. To determine the N_e with the theory of Oks, published data were fit to determine the N_e versus H_β width dependence [18]. Figure 17.3 shows the fitted H_β spectrum from the 10 μs delay along with its corresponding Stark width and electron number density. The noticeable dip in the residual in figure 17.3 near 486.5 nm is due to the subtraction of the fitted AlO spectra and represents the 1–1 vibrational peak from the transition of interest. Also

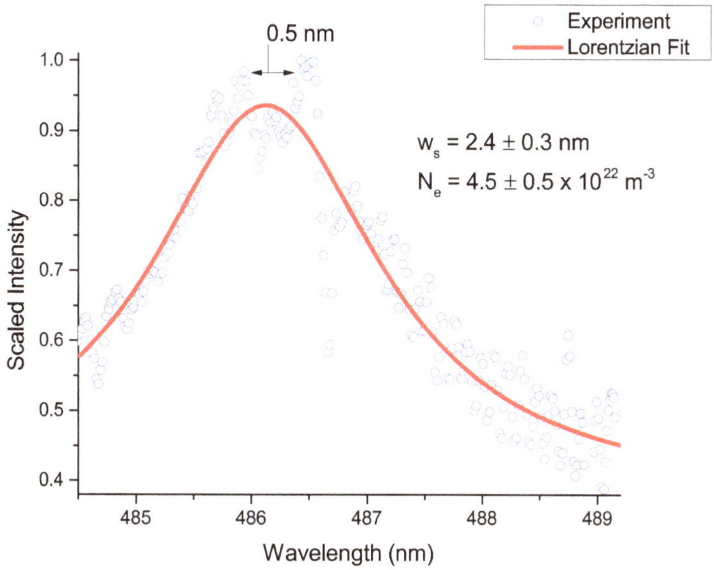

Figure 17.3. A fit of H_β superposition spectra at a 10 μs delay to a Lorentzian line profile. Reproduced from [2], © IOP Publishing Ltd. CC BY 3.0.

indicated in the figure is the 0.5 nm H_β peak separation. Though not considered in this work, this peak separation may also be used to infer the electron density [18]. The inferred N_e are found to be 4.5 ± 0.5, 3.8 ± 0.5 and 3.0 ± 0.6 × 1022 m^{-3}. Errors in the inferred N_e result from AlO superposition influences.

Accurate theoretical calculations of the diatomic, molecular AlO spectra have been instrumental in fitting to experimental data, particularly for studies with spectral resolutions of the order of 0.1 nm [14, 15]. The accuracy of such fitting is based, partially, in the ability to calculate the line strength tables with a resolution ranging from one to several orders of magnitude greater than that of the measured spectra. Recent efforts involving laser ablation experiments of an aluminum alloy target have been performed in order to accurately determine the temperature decay of aluminum containing laser-induced plasma using a line strength table with such accuracy. A summary of aluminum laser ablation efforts at UTSI may be found in references [1, 2, 9, 10, 19]. Figure 17.4 illustrates the $\Delta v = 0$ sequence at a spectral resolution of 0.09 nm and a corresponding, measured spectrum from two-dimensional slit height versus wavelength data. Representative results of these investigations conducted with a one dimensional array detector are presented below. Figure 17.5 shows measured and fitted AlO spectra [20], using the AlO line strength table, at two time delays following plasma initiation.

The objectives of the AlO work was to characterize the temporal temperature decay of laser-induced plasma near the surface of an aluminium target and was accomplished with time-resolved LIBS measurements. A Nelder–Mead fitting algorithm was employed to infer the temperature from the experimental spectra and errors were determined from spectral resolution and baseline offset variations. The temperatures were found to decrease as a function of time from 5173 ± 290 K to

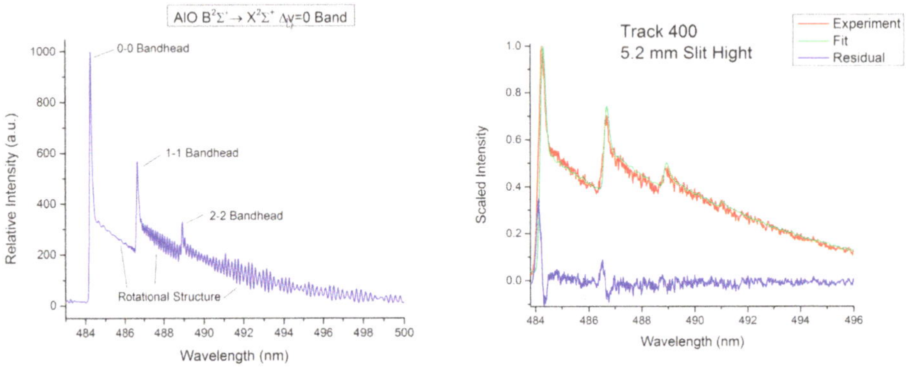

Figure 17.4. Calculated $T = 4$ kK, $\Delta\lambda = 0.09$ nm (left), measured and fitted (right) AlO emission spectra. Reprinted from [20] with permission of Elsevier.

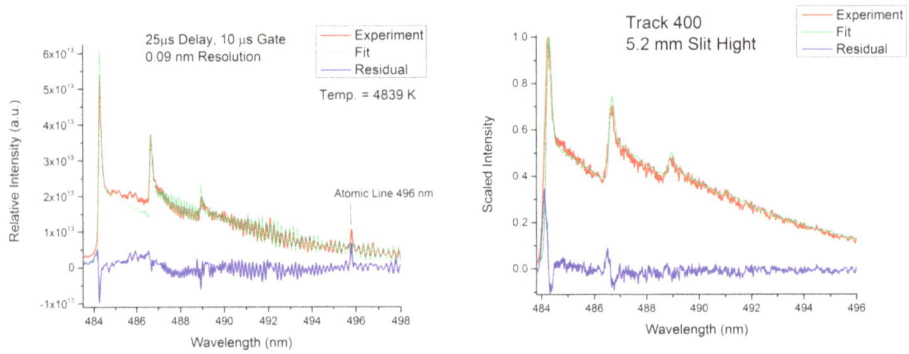

Figure 17.5. Fitted (green line) and measured (red line) AlO $B^2\Sigma^+ \longrightarrow X^2\Sigma^+$, $\Delta v = 0$ spectra collected at 25 µs (left) and 80 µs (right) following optically induced breakdown. The inferred temperatures are 4.839 and 3.699 kK, respectively. Reprinted from [20] with permission of Elsevier.

3862 ± 46 K from 10 to 100 µs time delays. The decay of the temperature was also found to slow at later time delays. Spatially resolved AlO measurements were used to infer the temperature along the height of the plasma and showed the temperature increases above the plasma edge. The electron density of the plasma was inferred from measurements of H_β spectra superimposed on AlO measurements at time delays preceding 20 µs and were found to decrease from 4.5 ± 0.5 to $3 \pm 0.5 \times 1016$ cm^{-3}. Observations of the slowing of the temperature decay, the temperature increase above the plasma, and H_β superposition spectra suggest that combustion occurs.

References

[1] Dors I G, Parigger C G and Lewis J W L 1998 *Opt. Lett.* **23** 1778
[2] Surmick D M and Parigger C G 2014 *J. Phys.: Conf. Ser.* **548** 012046

[3] Bransden B H and Joachain C J 2003 *Physics of Atoms and Molecules* 2nd edn (New York: Prentice-Hall)
[4] Piehler T N, DeLucia F C, Munson C A, Homan B E, Miziolek A W and McNesby K L 2005 *Appl. Opt.* **44** 3654
[5] De F C, Lucia, Harmon R S, McNesby K L, Winkel R J and Miziolek A W 2003 *Appl. Opt.* **42** 6148
[6] Sabsabi M and Cielo P 1995 *Appl. Spectrosc.* **44** 3654
[7] Rai A K, Yueh F and Signh J P 2003 *Appl. Opt.* **42** 2078
[8] Lightstone J M, Carney J R, Boswel C J and Wilikinson J 2007 *AIP Conf. Proc.* **955** 1255
[9] Parigger C G, Hornkohl J O and Nemes L 2007 *Appl. Opt.* **46** 4026
[10] Surmick D M and Parigger C G 2014 *Appl. Spectrosc.* **68** 992
[11] Parigger C G, Woods A C, Witte M J, Swafford L D and Surmick D M 2014 *J. Vis. Exp.* **84** 51250
[12] Nelder J A and Mead R 1965 *Comput. J.* **7** 308
[13] Lagarias J C, Reeds J A, Wright M H and Wright P E 1998 *SIAM J. Optim.* **9** 112
[14] Parigger C G and Hornkohl J O 2011 *Spectrochim. Acta Part A: Mol. Biomol. Spectrosc.* **81** 403
[15] Nemes L, Hornkohl J O and Parigger C G 2005 *Appl. Opt.* **44** 3686
[16] Konjević N, Ivković M and Sakan N 2012 *Spectrochim. Acta Part B: At. Spectrosc.* **76** 16
[17] Ispolatov Y and Oks E 1994 *J. Quant. Spectrosc. Radiat. Transfer* **51** 129
[18] Oks E, Parigger C G and Plemmons D H 2003 *Appl. Opt.* **42** 5992
[19] Parigger C G 2006 *AIP Conf. Proc.* **874** 101
[20] Parigger C G, Gautam G, Woods A C, Surmick D M and Hornkohl J O 2014 Trends *Appl. Spectrosc.* **11** 1

Chapter 18

Hydroxyl, OH

Ultraviolet OH radiation measurements and studies have been employed successfully to characterize atmospheric phenomena and combustion processes [1–5]. The use of the hydroxyl radical in combustion diagnostics has been reported frequently—OH is a prime molecule for the study of thermal and chemical phenomena [6–18]. The preferred techniques comprise laser-induced fluorescence, two-dimensional imaging possibly with single-pulse excitation, and degenerate four-wave mixing, for accurate measurement of temperature at various stages of a particular chemical process. This chapter discusses various aspects of OH measurements [19, 20] in nanosecond laser-induced breakdown spectroscopy.

The hydroxyl-radical is also a potential monitor of plasma temperature. Emission spectroscopy is primarily employed to determine the spectroscopic temperature of OH that is formed by recombination. In the characterization of laser-induced air plasma, particularly the $A^2\Sigma \leftrightarrow X^2\Pi$ UV system of OH is of interest. Applications of laser-induced breakdown plasma include the initiation of combustion, or laser-ignition. Prior to the use of OH as an indicator and monitor of laser-ignition, knowledge is required of the spectral signatures of the laser-induced breakdown plasma that can ignite combustible mixtures. Ignition delays in laser-induced combustion can be in the order of some 10 μs, yet air-plasma recombination spectra indicate a temperature of 3 kK and a number density of 3×10^{16} cm^{-3} at a time delay of 200 μs [20]. Hydroxyl is clearly detected in time-resolved spectral studies of laser-induced breakdown micro-plasma in SATP laboratory air in the wavelength range of 305 nm to 322 nm.

Laser-induced breakdown spectroscopy (LIBS) is utilized to characterize optical breakdown in SATP laboratory air. Optical breakdown is accomplished by focusing 1064nm laser radiation of 3.5 ns pulse width; intensities of typically 10 TW/cm^2 resulted in the focal volume. A Coherent Infinity 40-100 Nd:YAG laser was operated at a frequency of 10 or 100 Hz. The spectra were resolved with a 0.64-m

doi:10.1088/978-0-7503-1989-8ch18

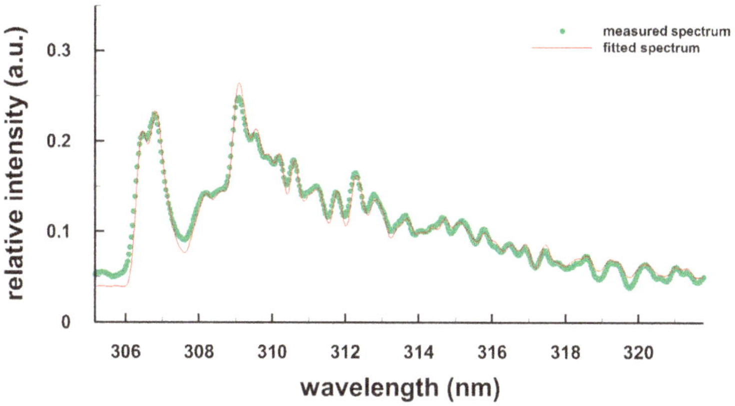

Figure 18.1. Measured and fitted air breakdown spectra, 60 μs time delay, $T = 4$ kK. Adapted with permission from [20]. © The Optical Society.

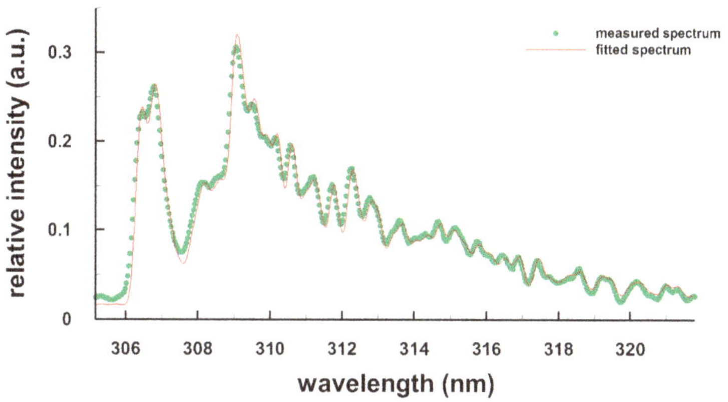

Figure 18.2. Measured and fitted air breakdown spectra, 100 μs time delay, $T = 3.1$ kK. Adapted with permission from [20] © The Optical Society.

Jobin-Yvon spectrometer and measured by the use of an intensified array detector and a multichannel analyzer. The gate-widths were 10 μs for data recorded between 20 μs and 50 μs after optical breakdown, 20 μs for data recorded at 60 μs and 80 μs delays, and 50 μs for data recorded at delays of 100 μs and 200 μs. Wavelength calibrations were performed with standard light sources and the sensitivity correction was accomplished with a deuterium lamp. The detector's data were corrected for dark-noise contribution.

Figures 18.1 and 18.2 show time-resolved recombination emission spectra that were measured at time delays of 60 and 100 μs, respectively, after the optical breakdown event in air. The spectral resolution amounted to 0.32 nm (32 cm^{-1}).

Predominantly the N_2 second positive system contributes to the spectral emissions at a delay time of some 10 μs, OH emissions can be recognized at delay times of 20 μs. The band structure of the N_2 second positive system is greatly reduced at a delay time of 40 μs, and subsequent emissions in the plasma decay are primarily from the $A^2\Sigma \leftrightarrow X^2\Pi$ UV system of OH in the wavelength region of 305–322 nm, but significant background radiation occurs at delay times of 60 μs (figure 18.1).

The spectral fitting of the emissions recorded at a delay time of 60 μs was further investigated. Figure 18.1 reflects results from the application of previously discussed algorithms [21–23] for the computation of diatomic line-strengths [24–26] that are constructed from OH spectral data bases [27–30]. The data reduction included a systematic variation of the background contributions which were computed by the use of the NEQAIR code. The calculated background was subtracted from the measured spectrum, subsequently, the Nelder–Mead [31, 32] least-square fitting algorithm was employed to determine the best-fit temperature. The background subtraction was accomplished by scaling experimental and synthetic spectra such that the integrated emissions are identical. For time delays of more than 150 μs, or for temperatures of approximately 3 kK, the background contribution is in the order of 1% or less of the integrated emission. Consequently, the temperature is exclusively found from the OH spectral distribution. The background radiation contains information of species number densities, however, for increased delay times or lower temperatures the error bars increase for the species number densities. Although NEQAIR wavelength positions are inadequate for precise fitting, the integrated emissions in the region of 305–322 nm are accurately predicted. Furthermore, the assumption of thermodynamic equilibrium in the OH spectra produced by optical breakdown of air is reasonable for time delays of the order of 100 μs.

References

[1] Levin D A, Laux C O and Kruger C H 1999 *J. Quant. Spectrosc. Radiat. Transfer* **61** 377

[2] Crosley D R and Jeffries J B 1996 Temperature measurements by laser-induced fluorescence of the hydroxyl radical *Temperature: Its Measurement and Control in Science and Industry* vol 6 ed L J F Schooley (New York: American Institute of Physics)

[3] Battles B E and Hanson R K 1995 *J. Quant. Spectrosc. Radiat. Transfer* **54** 521

[4] Levin D A, Laux C O and Kruger C H 1995 A general model for the spectral calculation of oh radiation in the ultraviolet *AIAA paper 95-1990 presented at the 26th AIAA Plasmadynamics and Lasers Conf. (San Diego, CA)*

[5] Laoux C O 1993 Optical diagnostics and radiative emission of air plasmas *PhD Dissertation* (Department of Mechanical Engineering, Stanford University)

[6] Rakestraw D J, Farrow R L and Dreier T 1990 *Opt. Lett.* **15** 709

[7] Kohse-Höinghaus K, Meier U and Attal-Tretout B 1990 *Appl. Opt.* **29** 1560

[8] Attal-Tretout B, Schmidt S C, Crete E, Dumas P and Taran J P 1990 *J. Quant. Spectrosc. Radiat. Transfer* **43** 351

[9] Desgroux P and Cottereau M J 1991 *Appl. Opt.* **30** 90

[10] Ewart P and Kaczmarek M 1991 *Appl. Opt.* **30** 3996–99

[11] Cignoli F, Benecchi S and Zizak G 1992 *Opt. Lett.* **17** 229

[12] Nyholm K, Maier R, Aminoff C G and Kaivola M 1993 *Appl. Opt.* **32** 919

[13] Nyholm K, Fritzon R and Alden M 1993 *Opt. Lett.* **18** 1672

[14] Seitzman J M, Hanson R K, Debarber P A and Hess C F 1994 *Appl. Opt.* **33** 4000

[15] Rahn L A and Brown M S 1994 *Opt. Lett.* **19** 1249
[16] Allen M G, McManus K R, Sonnenfroh D M and Paul P H 1995 *Appl. Opt.* **34** 6287
[17] Palmer J L and Hanson R K 1996 *Appl. Opt.* **35** 485
[18] Neuber A A, Janicka J and Hassel E P 1996 *Appl. Opt.* **35** 4033
[19] Parigger C G, Lewis J W L, Plemmons D P and Hornkohl J O 1996 Nitric oxide optical breakdown spectra and analysis by the use of the program NEQAIR *Laser Appl. Chem. Bio. Env. Anal.* **3** 85
[20] Parigger C G, Guan G and Hornkohl J O 2003 *Appl. Opt.* **42** 5986
[21] Tellinghuisen J 1972 *J. Mol. Spectrosc.* **44** 194
[22] Tellinghuisen J 1974 *J. Comput. Phys.* **6** 221
[23] Zare R N, Schmeltekopf A L, Harrop W J and Albritton D L 1973 *J. Mol. Spectrosc.* **46** 37
[24] Tatum J B 1967 The interpretation of intensities in diatomic molecular spectra *Astrophys. J.* **XIV** 21
[25] Hornkohl J O, Parigger C G and Lewis J W L 1991 *J. Quant. Spectrosc. Radiat. Transfer* **46** 405
[26] Hornkohl J O and Parigger C G 1996 *Am. J. Phys.* **64** 623
[27] Dieke G H and Crosswhite H M 1962 *J. Quant. Spectrosc. Radiat. Transfer* **2** 97
[28] Coxon J A 1980 *Can. J. Phys.* **58** 933
[29] Coxon J A and Foster S C 1981 *Can. J. Phys.* **60** 41
[30] Coxon J A, Sappey A D and Copeland R A 1991 *J. Mol. Spectrosc.* **145** 41
[31] Nelder J A and Mead R 1965 *Comput. J.* **7** 308
[32] Salieri F, Quarteroni A and Sacco R 2000 *Numerical Mathematics* (New York: Springer)

Chapter 19

Titanium monoxide, TiO

Various research efforts regarding laser-induced breakdown spectroscopy (LIBS) of titanium normally focus on the atomic and ionic Ti spectral transition lines. However, after a characteristic time subsequent to laser ablation, these lines are no longer discernable. During this temporal window, the diatomic molecular transition lines of titanium monoxide (TiO) are prominent in the laser-induced plasma (LIP) emissions. This chapter discusses recent TiO research [1] that was also presented at a recent line shape conference [2].

The molecule TiO has long been studied in the contexts of stellar emissions, allowing for some of the molecular transition bands to be accurately computed from theory. In this research, optical emission spectroscopy (OES) of laser-induced plasma (LIP) generated by laser ablation of titanium is performed in order to infer temperature as a function of time subsequent to plasma formation. The emission spectra of the resulting ablation plume are imaged as a function of height above the sample surface. Temperatures are inferred over time delays following plasma formation ranging from 20 to 200 μs. Computed TiO $A^3\Phi - X^3\Delta$, $\Delta v = 0$ transition lines are fit to spectral measurements in order to infer temperature. At $t_{\mathrm{delay}} = 20$–80 μs, the observed plume contains two luminescent regions each with a distinctly different temperature. As the plume evolves in time, the two regions combine and an overall temperature increase is observed.

19.1 Introduction

Laser-induced breakdown spectroscopy (LIBS), as an experimental technique, has grown in popularity in large part due to the simplicity of the required equipment. The traditional LIBS apparatus consists of a single laser, serving as an excitation source, and a spectrometer equipped with a detector for gathering spectral emissions [3]. The success and growth of future LIBS applications require a deeper

doi:10.1088/978-0-7503-1989-8ch19

understanding of the fundamental processes occurring as a plasma is observed at various time delays, given various experimental conditions [4]. To this extent, much of the ongoing fundamental research utilizing LIBS seeks to observe and analyze specific atomic and molecular species present in the laser-induced plasma (LIP) [5–9]. By closely examining the properties of spectra from specific atoms and molecules at various regions of the plasma, as it evolves in time, a description of the characteristic processes going on throughout the plasma can be obtained. Of particular interest is the time domain for which molecular recombination overcomes electron collisions as the dominant mechanism concerning plasma emissions.

As LIP evolves in time, the characteristic emitting species dominating the spectra will evolve as well. An initial LIBS spectrum, typically within the first tens of nanoseconds, is dominated by a spectral continuum contribution. Mechanisms, such as inverse bremsstrahlung and radiative recombination involving free electrons, are responsible for this spectral continuum. As the plasma expands and cools, ionic spectral line transitions appear in the plasma emissions, superposed to the continuum contribution. The more highly ionized species will become prominent first, followed by the lesser ionized species and ultimately the neutral atomic transition lines [10–12]. Three-body recombination occurring with plasma expansion causes a rapid decrease of the electron density in the plasma [13]. As a consequence of the decreasing electron density in the plasma, each emission line will become progressively narrower during the plasma expansion [14].

After a characteristic time, depending on electron density, the elements in the plasma and the plasma expansion, emissions from radiative molecular transitions will become discernible as the prominence of atomic spectral lines observed in the plasma decreases [15]. Of particular interest for this research is the diatomic molecular radiative transitions of the plasma. Typically, this phenomenon is attributed to the recombination of atoms to form molecules. However, the detection of diatomic transitions of certain species at far earlier stages of plasma evolution when compared to other diatomic transitions in similar conditions leads to the conclusion that not all diatomic transitions are a product of recombination. In these scenarios, it is hypothesized that the molecule was present prior to plasma formation. However, many diatomic molecules observed in laser-induced plasma are unstable in typical laboratory environments. For such molecules, LIBS offers a suitable method for observation. The spectral transitions of such molecules can be collected typically for tens of microseconds. Often times, these emissions are quite luminous, from the ultra-violet to the infrared.

For the purposes of analyzing the laser ablation plasma above the Ti surface at time delays on the order of tens to hundreds of microseconds, TiO spectral transitions are utilized to infer temperature. This requires the computation of new, accurate line-strength tables. Diatomic quantum theory is utilized in computing line-strengths of spectral transitions [6, 16]. The specifics concerning the computation of the line-strengths used in this research are presented elsewhere [1, 17]. Given a few parameters, such as resolution and temperature, the spectral line-strengths can be used to generate a synthetic spectrum representative of the radiant transitions of a diatomic molecule. In order to infer micro-plasma parameters, the experimentally collected spectra believed to contain the diatomic spectral signatures are fit with the computed spectra for the molecule [1, 17, 18].

Previous studies observed the TiO $B^3\Pi \rightarrow X^3\Delta$ and $A^3\Phi \rightarrow X^3\Delta$, $\Delta v = 0$ transition bands in the ablation plasma [1, 19]. The collected spectra were then fit with computed spectra in order to infer temperature as a function of delay time. Temperature inferences, resulting from the use of an intensified linear diode array arrangement to gather time-resolved spectra, typically contained a local minimum occurring between a 40 and 60 μs delay time. At 60 μs, when the expanding plasma is thought to be cooling, this unexpected result indicated an increase in temperature. In order to further investigate the situation, the current research observes the TiO $A^3\Phi - X^3\Delta$, $\Delta v = 0$ transition band at time delays ranging from 20 to 200 μs. Utilizing a gated two-dimensional ICCD detector, the gathered emissions are time-resolved and imaged as a function of height above an ablation surface. The gathered spectra are then fit with computed spectra representing the TiO $A^3\Phi - X^3\Delta$ band to infer temperature.

19.2 Experiment

A Ti 6-4 sample is ablated by radiation from a Nd:YAG laser with 160 mJ per pulse and a nominal pulse width of 13 ns. The laser radiation is supplied at 10 Hz and focused to approximately a 1 mm spot size. The beam propagates vertically downward onto the flat Ti surface such that the beam path is parallel to the slit aperture of a Jobin-Yvon spectrometer. The resulting ablation plume emissions are imaged onto the slit in a 1:1 manner. The spectrometer is equipped with an Andor iStar ICCD camera, providing both temporal and spatial resolution. In order to enhance the signal-to-noise ratio, the 1024 × 1024 pixels provided by the detector are consolidated into 32 spectra (1024 pixels across) gathered along the height of the plume for each image. Additionally, 200 accumulations are used to produce each image. At $t_{\text{delay}} = 20$–200 μs, spectra in the range of 705–715 nm are collected every 20 μs utilizing a 2 μs gate width. In this spectral region, at these time delays, TiO transition lines are collected. The gathered spectra are then fit with computed spectra from the TiO $A^3\Phi - X^3\Delta$, $\Delta v = 0$ transition band in order to infer temperature.

19.3 Results

Figure 19.1 provides an example of an experimentally obtained spectrum fit with a computed spectrum for the TiO γ transition. The collected spectrum of figure 19.1 represents a single horizontal slice of a larger image. Figure 19.2 is an example of such an image. The plot just below the image in figure 19.2 represents the vertical sum of the pixels in the image. Such a summation is likened to using a linear diode array detector. While spectra computed from theory are used to infer the TiO transition temperature along the height of the image, the spectrum obtained by integrating the height of the detector can be similarly analyzed. The result is typically a temperature corresponding to the inferred temperature at the hottest region of the image.

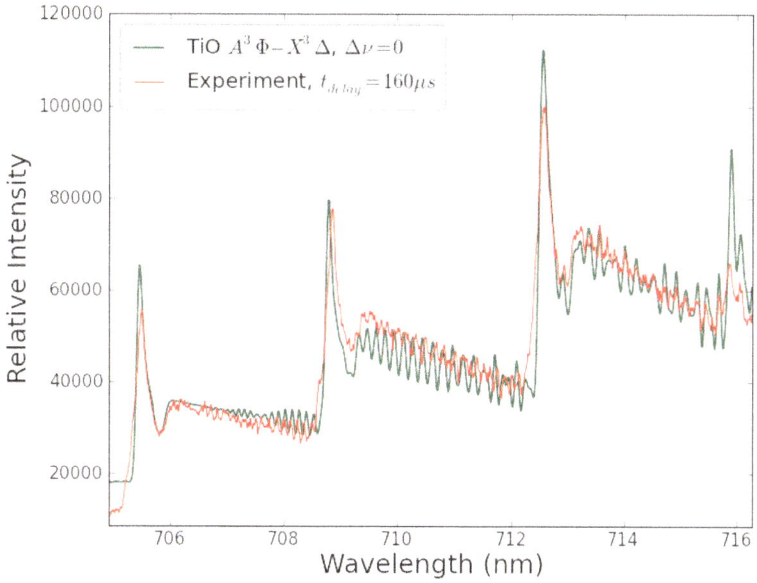

Figure 19.1. This spectrum represents a horizontal slice of one the gathered images fit with computed spectra corresponding to $T = 3400$ K. The measured spectrum was collected at $t_{\text{delay}} = 160$ μs. Reproduced from [2]. © IOP Publishing Ltd. CC BY 3.0.

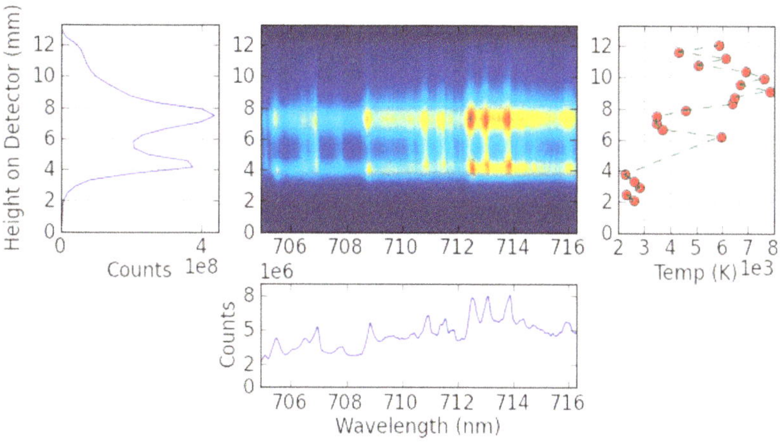

Figure 19.2. Recorded spectra of laser ablation plasma above a titanium surface recorded at $t_{\text{delay}} = 40$ μs with a 2 μs gate width. Reproduced from [2]. © IOP Publishing Ltd. CC BY 3.0.

By studying figures 19.2–19.6, some very interesting phenomena becomes apparent. The temperature inferred from the TiO molecular transitions increases with height along the ablation plume. This phenomena is consistent throughout $t_{\text{delay}} = 20$–200 μs. More interestingly, it is clear that between $t_{\text{delay}} = 20$–60 μs there are two distinct regions of the plasma. One region rests several millimeters above the sample surface, while the other appears to rise from the surface before the two

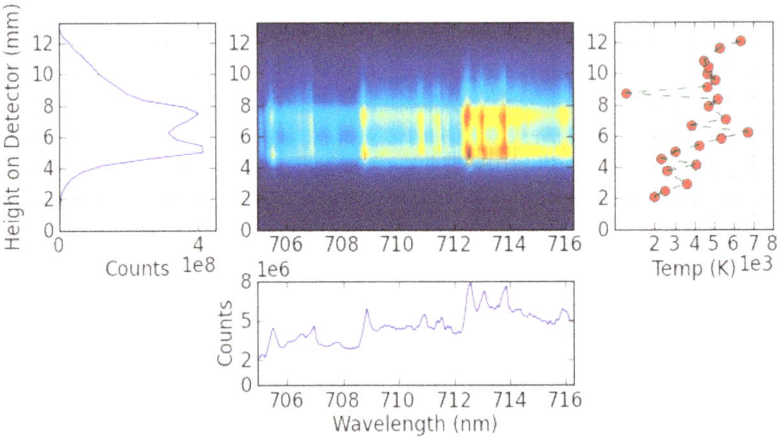

Figure 19.3. Recorded spectra of laser ablation plasma above a titanium surface recorded at $t_{\mathrm{delay}} = 60$ μs with a 2 μs gate width. Reproduced from [2].

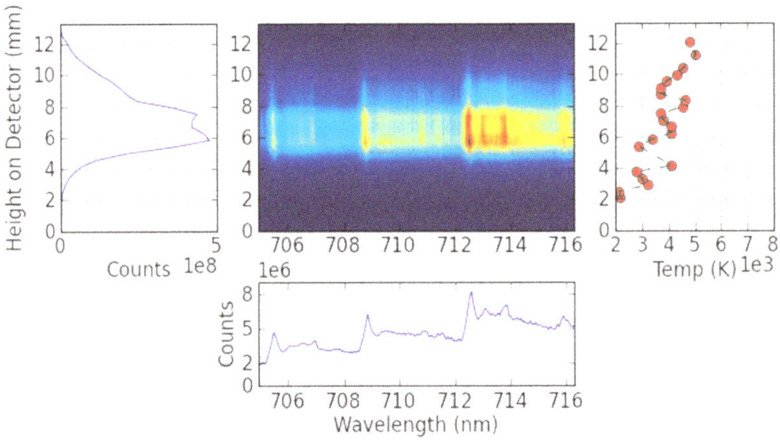

Figure 19.4. Recorded spectra of laser ablation plasma above a titanium surface recorded at $t_{\mathrm{delay}} = 80$ μs with a 2 μs gate width. Reproduced from [2].

ultimately combine. After the two regions combine (~100 μs), there is a noticeable bulge in the uppermost part of the intensity profile.

When the temperature inferences made from emissions of the two regions are compared, it is apparent that the inferred temperature of the TiO emissions from the region closer to the sample surface is distinctly lower than that of the region further above the surface. This is most clearly evident in figures 19.2 and 19.3. Particularly when examining the inferred temperatures of figure 19.2, it is important to keep in mind that we are struggling to make temperature inferences resulting from fitting computed TiO spectra to the collected spectra on account of signal-to-noise issues. Notice at the earlier time delays, the characteristic structure of the TiO $A^3\Phi - X^3\Delta$ band is hardly discernable. Much of the spectral signatures observed in this time

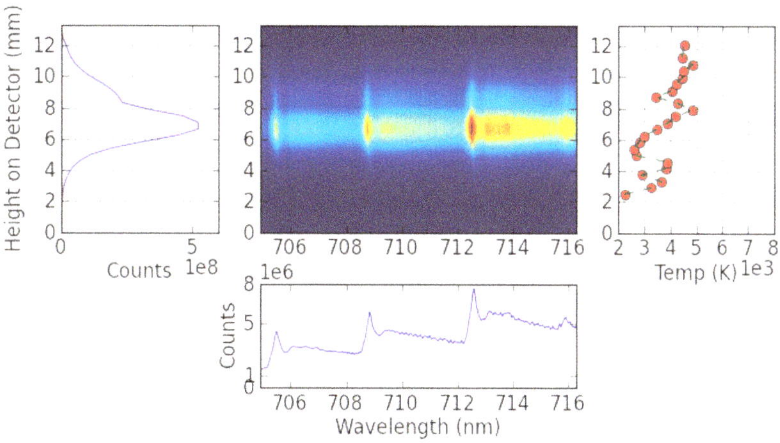

Figure 19.5. Recorded spectra of laser ablation plasma above a titanium surface recorded at $t_{\text{delay}} = 100$ μs with a 2 μs gate width. Reproduced from [2]. © IOP Publishing Ltd. CC BY 3.0.

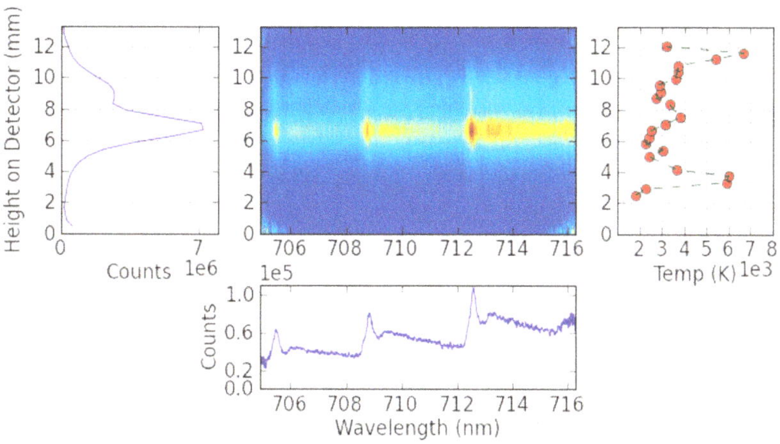

Figure 19.6. Recorded spectra of laser ablation plasma above a titanium surface recorded at $t_{\text{delay}} = 200$ μs with a 2 μs gate width. Reproduced from [2]. © IOP Publishing Ltd. CC BY 3.0.

domain are atomic line shapes. When inspecting the inferred temperature, more confidence is given to those values inferred from the brighter regions of the plume. These regions are made brighter by the luminescent TiO transitions. This conveniently allows for succinct comparison between the two brightest regions of figure 19.2, as the TiO spectral signatures in the dimmer regions are less discernable from the other spectral contributions.

Examining the plots to the left of the images in figures 19.2–19.6 provides insight into the luminosity along the height of the plasma. The motion of the lower peak toward the upper peak can be used as a measure of the kinetics of the luminous lower region of the plasma. These plots for figures 19.2–19.4 also illustrate an increase in luminosity of the lower region as it approaches the upper. At later time

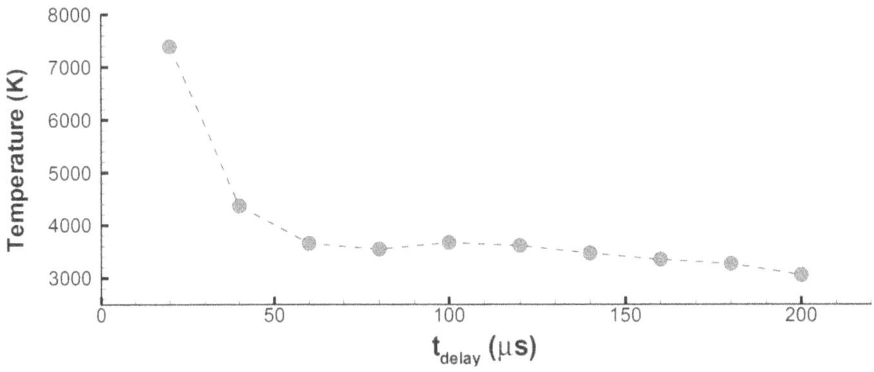

Figure 19.7. Temperature as a function of time delay inferred from the analysis of the spectra resulting from integrating along the height of the detector. Reproduced from [2].

delays, t_{delay} = 100–200 μs, a hump is observed at the top luminous region of the ablation plume. The height of the plasma corresponding to the slight dip below the hump is determined to be hotter than the regions above or below the dip. While the hump appears to be a consequence of a snowplough effect at the top of the plasma, two-dimensional spatial imaging of the plume would possibly provide an explanation. The current apparatus observes primarily along the axial distance perpendicular to the surface; however, imaging in the radial direction would likely provide a crown-shaped profile, as in figure 8 of Aguilera *et al* [20].

Comparison of the inferred temperature from the upper region of figure 19.2 shows remarkable agreement with the results obtained previously at this same time delay (t_{delay} = 40 μs) using this same window (2 μs) [19]. If one makes the same comparison with the upper region of figure 19.3 and Woods *et al* [19] at t_{delay} = 60 μs, there is some agreement for the inferred temperatures here as well. When the spectra obtained by integrating over the height of the images are analyzed, the inferred temperatures typically correspond to a temperature slightly elevated from that of the upper luminescent region. Figure 19.7 illustrates temperature versus time delay.

Examining figure 19.7, a local temperature increase of ≈100 K occurs at t_{delay} = 100 μs. Previous investigations concerning both the TiO $B^3\Pi \rightarrow X^3\Delta$ and $A^3\Phi - X^3\Delta$ systems have also found an unexpected temperature increase at later time delays [1, 19]. Inspection of the inferred temperatures along the height of the image over t_{delay} = 40–100 μs indicate that the lower region undergoes drastic heating, as the cooler TiO molecules of the lower luminescent region interact with the warmer TiO molecules of the upper region. This interaction results in an overall increase in TiO temperature in the plasma.

While more investigations are required to fully characterize this bright region rising from the surface, some hypotheses can be formed. By the ability to fit computed TiO spectra to this emitting region, it is determined that the TiO molecule is the dominant emitter. In the upper region, the TiO transitions are likely the result of recombination. This is consistent with the delay time in which the transitions become discernible given the excitation source and the evolution of the temperature and electron density, as determined by the measurements made within the first hundreds of nanoseconds. The drastic temperature difference at $t_{delay} = 40$ μs between the upper and lower luminescent regions remains largely unexplained. As the cooler plasma temperatures are typically found to be in the lower regions, perhaps a sudden flurry of recombination occurs near the sample surface, overtaking electron impacts as the driving mechanism for emissions. This trend then rises along the plasma until it reaches the upper region. However, because the temperatures inferred from the TiO transitions in the lower region are starkly less than those of the upper region and gradually increase as the regions interact, an alternative theory would maintain that the mechanism causing the radiative transitions of the diatomic molecule is different between the two regions. During laser ablation at this intensity, the sample surface will become molten for a characteristic duration [21]. Considering the lower region is much cooler and appears to heat as it leaves the surface, possibly the TiO emissions in this region are emanating from bulk titanium being ejected from the sample surface.

References

[1] Woods A C, Parigger C G and Hornkohl J O 2012 *Opt. Lett.* **37** 5139

[2] Woods A C and Parigger C G 2014 *J. Phys.: Conf. Ser.* **548** 012037

[3] Parigger C G, Woods A C, Witte M J, Swafford L D and Surmick D M 2014 *J. Vis. Exp.* **84** 51250

[4] Amoruso S, Armenante M, Berardi V, Bruzzese R, Spinelli N and Velotta R 1997 *J. Phys.* B **36** 3227

[5] Parigger C G and Oks E 2010 *Int. Rev. At. Mol. Phys.* **1** 13

[6] Hornkohl J O, Nemes L and Parigger C G 2009 Spectroscopy, dynamics and molecular theory of carbon plasmas and vapors *Advances in the Understanding of the Most Complex High-Temperature Elemental System* ed L Nemes and S Irle (Singapore: World Scientific) ch 4

[7] Witte M J, Parigger C G, Bullock N A, Merten J A and Allen S D 2014 *Appl. Spectrosc.* **68** 367

[8] Parigger C G, Swafford L D, Woods A C, Surmick D M and Witte M J 2014 *Spectrochim. Acta Part B: At. Spectrosc.* **99** 28

[9] Parigger C G, Woods A C, Surmick D M, Swafford L D and Witte M J 2014 *Spectrochim. Acta Part B: At. Spectrosc.* **99** 15

[10] Albert O, Roger S, Glinec Y, Loulergue J C, Etchepare J, Boulmer-Leborgne C, Perriére J and Millon E 2003 *Appl. Phys.* A **76** 319

[11] De Giacomo A 2003 *Spectrochim. Acta Part B: At. Spectrosc.* **58** 71

[12] Capitelli M 2004 *Spectrochim. Acta Part B: At. Spectrosc.* **59** 271

[13] Casavola A, Colonna G and Capitelli M 2003 *Appl. Surf. Sci.* **208** 85

[14] Griem H 1974 *Spectral Line Broadening by Plasma* (New York: Academic)

[15] De Giacomo A, Dell'Aglio M, Santagata A and Teghil R 2005 *Spectrochim. Acta Part B: At. Spectrosc.* **60** 935

[16] Parigger C G and Hornkohl J O 2011 *Spectrochim. Acta Part A: Mol. Biomol. Spectrosc.* **81** 403

[17] Parigger C G, Woods A C, Keszler A, Nemes L and Hornkohl J O 2012 *AIP Conf. Proc.* **1464** 628

[18] Hermann J, Perrone A and Dutouquet C 2001 *J. Phys.* B **34** 153
[19] Woods A C, Parigger C G and Hornkohl J O 2012 *Int. Rev. At. Mol. Phys.* **3** 103
[20] Aguilera J A and Aragón C 2004 *Spectrochim. Acta Part A: At. Spectrosc.* **59** 1861
[21] Bogaerts A and Chen Z 2005 *Spectrochim. Acta Part A: At. Spectrosc.* **60** 1280

Chapter 20

Nitric oxide, NO

Nitric oxide spectra are studied with laser-induced breakdown spectroscopy in laboratory (LIBS) air and in a 1:1 mole ratio N_2:O_2 gaseous mixture. This chapter summaizes laser-induced breakdown spectroscopy of NO that was recently presented at a line shape conference [1]. The temporally-resolved excitation temperatures are determined by fitting recorded spectra with predicted diatomic spectra [1] that are constructed from diatomic line strengths. For time delays of 25 and 50 μs, the fitted temperatures are 6800 and 6100 K, respectively. Analysis of experiments in the 1:1 N_2:O_2 mixture is also accomplished with the non-equilibrium air radiation (NEQAIR) program [2].

Optical breakdown plasmas are generated by focused Nd:YAG IR laser radiation of 1–100 TW/cm^2 intensity. Typically rich emission spectra result. The decaying laser-induced plasma can be characterized by the application of time-resolved spectroscopy techniques. Primarily recombination spectra occur subsequent to optical breakdown [3–5]. It is of interest to obtain temperature and species density information from such spectra.

In the near-UV wavelength region of 205–300 nm, the emission spectra are comprised of overlapped electronic transitions of primarily NO and N_2 and to a lesser degree of O_2 and atomic species. The spectral analysis of these superimposed spectra of such multiple species is most challenging, and the computer code for non-equilibrium air radiation (NEQAIR) is applied in the spectral analysis of the transient micro-plasma.

The program NEQAIR [6, 7] is widely used to predict and interpret optical spectra from air plasmas. The majority of NEQAIR applications concern atmospheric radiation, but the program is also applied in simulation facilities and has other applications such as the prediction of laser sustained plasmas. In this paper we apply the non-equilibrium air radiation code, in the form of NEQAIR8 [8, 9], to our knowledge for the first time, in the analysis of time-dependent optical breakdown spectra.

The analysis of the recorded experimental spectra from air breakdown is based on the use of calculated number densities of a dry air plasma composition at a pressure of 100 kPa [10]. With the assumption of local thermodynamic equilibrium, a one-parameter least-square fitting procedure is applied to find the spectroscopic temperature.

One of the many applications of LIBS is its use in diagnostics such as combustion studies. Quantitative analysis efforts include investigations of both atomic and molecular spectra [11]. In such works diatomic species of interest were CN, N_2, and NH [12, 13]. Of particular interest to combustion are quantitative calculations and analysis of nitric oxide (NO) spectra for studies of nitrogen containing flames and exhaust plumes of rockets. Considered are the near-UV spectra of the NO γ-band in the 140–340 nm range [14].

In order to quantitatively study the emission of diatomic molecules, diatomic line strengths for NO are utilized. Detailed aspects of NO computations are deferred to appendix D. The use of line strengths is effective for the calculation of diatomic, molecular spectra due to its symmetric nature [15]. In this chapter, the line-strength approach is briefly reviewed. The formal definition of the line strength is the total sum of degenerate states that produce the same spectral line

$$S_{\mathrm{ul}} = \sum_{\mathrm{u}} \sum_{\mathrm{l}} \left| \left\langle \mathrm{u} \right| T_k^{(q)} \left| \mathrm{l} \right\rangle \right|^2 \tag{20.1}$$

where u and l represent the upper and lower states and $T_k^{(q)}$ is the kth component of the irreducible tensor operator of rank q that gives rise to the spectral transition. This operator is the electric dipole operator and the line strength may be expressed as

$$S_{\mathrm{ul}} = S(n'v', nv)S(J', J) \tag{20.2}$$

where $S(n'v', nv)$ is the electronic–vibration transition strength and $S(J', J)$ is the Hönl–London line-strength factor. Expressing the line strength as in equation (20.2) allows for a direct analytical calculation of the line strength, and in turn, the intensity of a diatomic emission spectra which may be expressed as

$$I_{\mathrm{ul}} = \frac{16\pi^3 c(a_0 e)^2 C_{\mathrm{abs}} N_0}{3\epsilon_0 Q} C_\nu \tilde{\nu}_{\mathrm{ul}}{}^4 S_{\mathrm{ul}} e^{-hF_u/k_{\mathrm{B}}T}, \tag{20.3}$$

where a_0 is the Bohr radius, e is the electron charge, c is the speed of light, Q is the partition function, N_0 is the total population of the species, C_{abs} and C_ν are the absolute and relative calibration factors, h is Planck's constant, F_u is the upper term value, k_{B} is Boltzmann's constant, and T is the temperature. A simple, yet rigorous selection rule results from the expression in equation (20.2). Given the relative ease in calculating the Hönl–London line strength and the ability for the Hönl–London line strength to be zero, allowed transitions are those with a non-vanishing line strength. In this study, we apply this method to infer the emission temperature of the

NO emission spectra from measurements made in the nominal 230–240 nm range for detailed studies and in the range of 210–300 nm for overview investigations.

20.1 Experimental details

The measurement of NO breakdown spectra, laser-induced plasma is generated in standard ambient temperature and pressure (SATP) laboratory air. Optical breakdown is initiated by focussing 1064 nm, 12 ns pulsed laser radiation from a Q-switch, Nd:YAG laser operating in its fundamental mode with an average energy per pulse of 190 mJ. Light from the plasma emission is dispersed by a 0.64 m Jobin-Yvon czerny turner spectrometer equipped with 3600 grooves/mm grating. Spectra are recorded with an intensified, linear diode array coupled to an optical multichannel analyzer. Time synchronization for the laser pulse firing rate and the measurement rate of the linear diode array is achieved with the use of waveform, pulse, and delay generators. Exact experimental details and a video [16] summarize the timing apparatus used in this experiment and in other, similar laser-induced plasma studies involving atomic and diatomic spectra. Prior to analysis, all spectra are properly calibrated for the detector background and sensitivity response.

20.2 Results

The NO spectra are fit to theoretical calculations tabulated from the line strength method. Comparisons between experiment and theory are made using a Nelder–Mead algorithm, which is used for its ability to use fit multiple parameters simultaneously, including a variable baseline offset. The Nelder–Mead minimizes parameters by iteratively reducing the size of a geometric simplex where the number of vertices in the simplex is one more than the number of parameters to be fit [17, 18]. The parameters considered while fitting are the temperature, spectral resolution, and offset. The spectral resolution parameter is held constant for each fit and is determined from the physical limitations of the experimental apparatus. Given the wavelength region and the high groove density of the grating used, the spectral resolution of the experimental spectra was 0.056 nm which also accounts for slit width effects. The constant offset parameter was allowed to vary. NO spectra were collected and analyzed for 25, 50, and 75 μs time delays for air breakdown experiments and the inferred temperatures were found to be 6.8, 6.1, and 5.8 kK for 25, 50, and 75 μs time delays, respectively. The fitting results for the 25, 50, and 75 μs data are depicted in figure 20.1, 20.2, and 20.3 . As can be seen in the fits shown in figures 20.1 and 20.2, the diatomic line strength method has the capability to calculate and fit spectra that include many rotational lines.

20.3 Comparison with overview spectra

The recorded spectra and inferred temperatures versus time delay in the spectral 230–240 nm agree with previous overview spectra recorded in NO LIBS experiments in SATP air. Furthermore, the NO LIBS spectra are analogous to those measured in plasma torches [19]. In these overview experiments, laser-induced optical breakdown was accomplished by the focusing of the Nd:YAG 1064nm radiation of 3.5 ns pulse

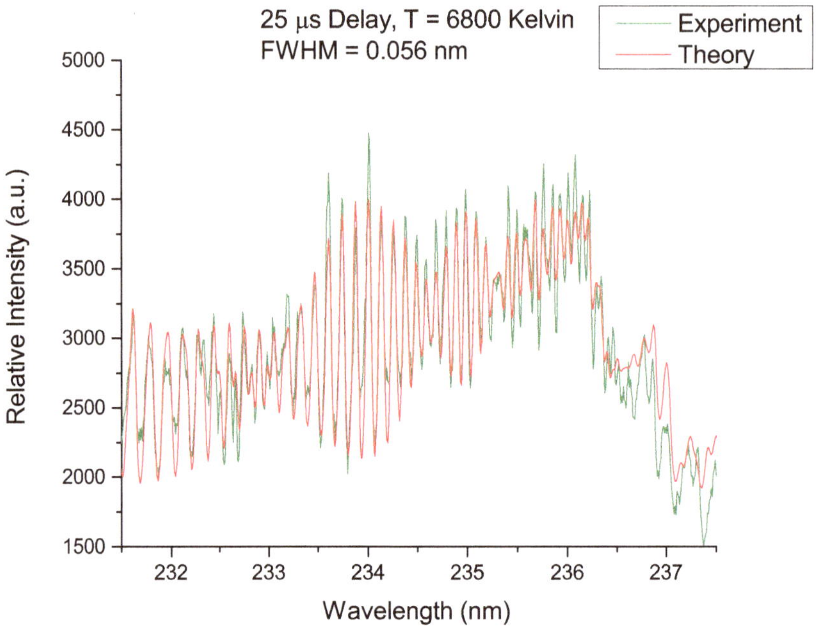

Figure 20.1. Fit to NO laser-induced plasma spectra at 25 μs time delay, T = 6.8 kK. Reproduced from [1]. © IOP Publishing Ltd. CC BY 3.0.

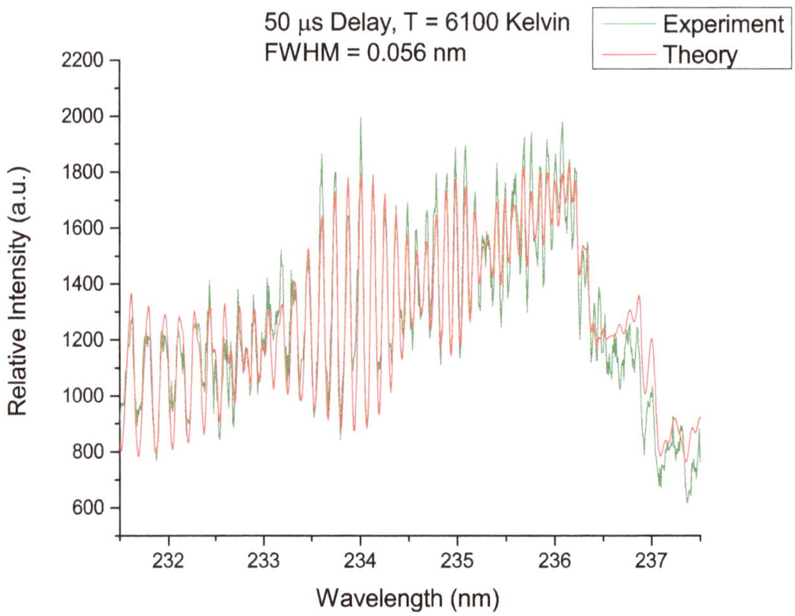

Figure 20.2. Fit to NO laser-induced plasma spectra at 50 μs time delay, T = 6.1 kK. Reproduced from [1]. © IOP Publishing Ltd. CC BY 3.0.

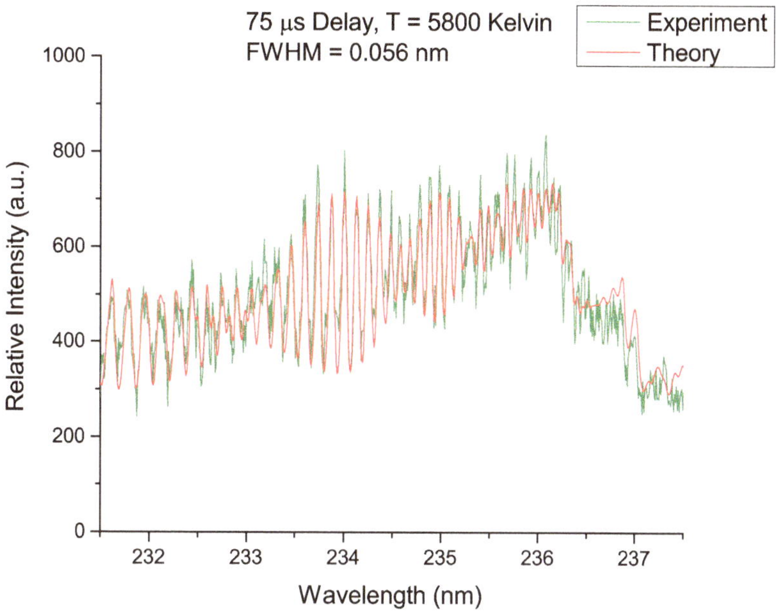

Figure 20.3. Fits to NO laser-induced plasma spectra at 75 μs time delay, T = 5.8 kK. Reproduced from [1].

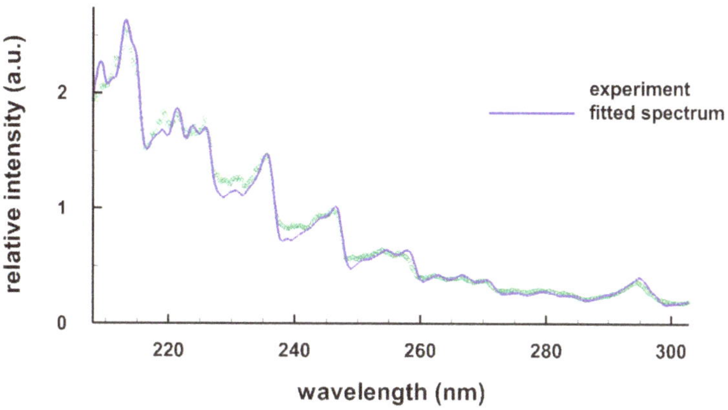

Figure 20.4. NO optical breakdown spectrum, 25 μs time delay, T = 6.7 kK [2].

width. A coherent infinity 40–100 Nd:YAG laser was operated at a frequency of 10 or 100 Hz. Optical breakdown was generated in SATP air and in laboratory cell that contained a 1:1 mole ratio nitrogen–oxygen mixture. The spectra were resolved with a 0.275-m Jarrel-Ash. Wavelength calibrations were performed with standard light sources, the sensitivity correction was accomplished with a deuterium lamp. The data were detector's dark-noise background subtracted.

Figure 20.4 and 20.5 show the experimental spectra that were recorded at a time delay of 25 μs and 50 μs, respectively, from the optical breakdown event in air they also include the fitted low-resolution NO spectra.

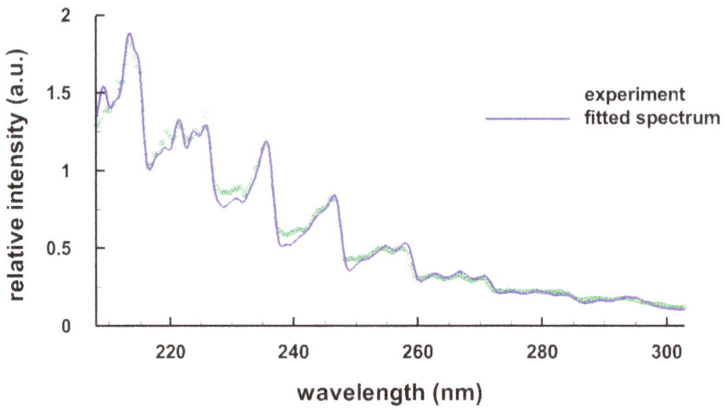

Figure 20.5. NO optical breakdown spectrum, 50 μs time delay, $T = 6\,\text{kK}$ [2].

Table 20.1. Neutral species number densities (cc^{-1}) for $T = 6000$ K and $T = 6700$ K.

T (K)	N	O	Ar	N_2	O_2	NO
6000	2.3×10^{17}	3.9×10^{17}	8.5×10^{15}	6.6×10^{17}	3.3×10^{14}	1.0×10^{16}
6700	4.6×10^{17}	3.1×10^{17}	6.8×10^{15}	3.6×10^{17}	7.8×10^{13}	4.5×10^{15}

Table 20.2. Ionized species and electron densities (cc^{-1}) for $T = 6000$ K and $T = 6700$ K.

T (K)	N^+	O^+	Ar^+	N_2^+	O_2^+	NO^+	e
6000	0	0	0	0	0	2.7×10^{14}	2.7×10^{14}
6700	4.7×10^{13}	4.8×10^{13}	0	0	0	4.4×10^{14}	5.4×10^{14}

Dry air is considered to be a mixture of nitrogen, oxygen and argon at room temperature, with all other species neglected. Starting from one mole of air at room temperature, the number densities of the species N, O, Ar, N^+, O^+, Ar^+, N_2, N_2^+, O_2, O_2^+, NO, NO^+, and the free electrons e, can be computed as a function of temperature with the chemical equilibrium and appliactions (CEA) program. The data from CEA serve as input for computation of synthetic spectra with the NEQAIR program, i.e., a Boltzmann distribution with equal translational, rotational, vibrational, and electron temperatures is assumed for the analysis presented here.

Tables 20.1 and 20.2 summarize the temperature and number densities that were utilized as input for the computer code NEQAIR8.

The spectra from the experiments that work were analyzed with the use of the non-equilibrium air radiation (NEQAIR) program and the chemical equilibrium (CEA) code. Temperatures, T, of 6700 and 6000 K are determined [2, 6, 12] for 25 and 50 μs time delays, respectively. It is important to note, however, that the approximate 1 nm spectral resolution is significantly lower than that presented in figure 20.1 and 20.2. The line strength method for calculating diatomic spectra provides a viable option for quantifying diatomic spectra. Furthermore, the method is useful for combustion studies that employ laser-induced florescence as a diagnostic method with spectra that often have comb-like structures. In such spectra, the highly resolved nature of a spectra calculated from the line strength is applicable [20]. This is found to be particularly useful in studies of combustion with high temperature flames [21].

The results indicate support for the program NEQAIR for applications in combustion [22–24] and plasma diagnostics. The 'background' contributions appear to be predicted well for selected diatomic transitions, and the synthetic NEQAIR spectra are particularly useful for the analysis of spectral studies with low spectral resolutions. Of course, non-equilibrium computations are of interest, e.g., for

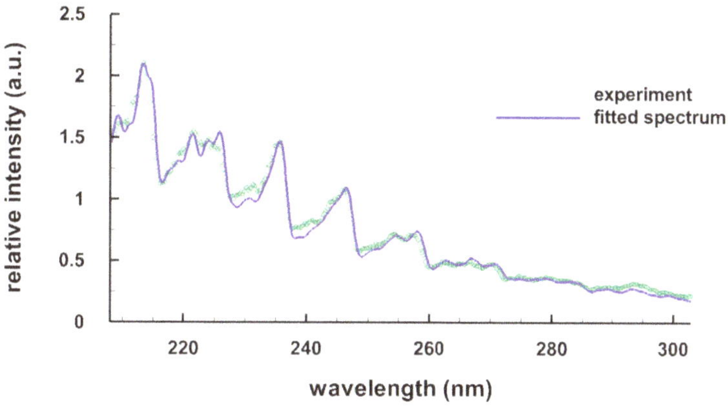

Figure 20.6. N_2:O_2 optical breakdown spectrum, 20 μs time delay.

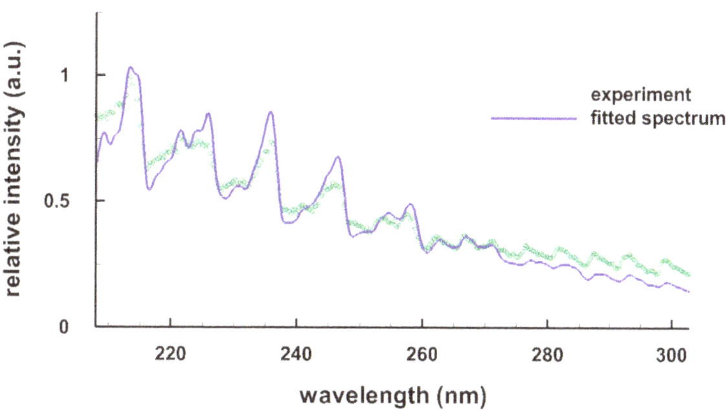

Figure 20.7. N_2:O_2 optical breakdown spectrum, 50 μs time delay. The spectra in the range of 210–260 nm are primarily due to NO, between 260 and 300 nm due to O_2^+.

different rotational and vibrational temperatures [25], but here we restricted our analysis to equilibrium distributions.

Further measurements investigated NO spectra from optical breakdown in 1:1 mole ratio mixture of N_2 and O_2 gas at a pressure slightly above 1 atm, but otherwise keeping the same experimental arrangement as for the SATP air studies. Figures 20.6 and 20.7 show the experimental spectra that were recorded at a time delay of 20 μs and 50 μs, respectively, from the optical breakdown events in the mixture. Only NO bands are included in the fitting, and O_2^+ spectral data would need to be included for a time delay of 50 μs.

References

[1] Hornkohl J O, Fleischmann J P, Surmick D M, Witte M J, Swafford L D, Woods A C and Parigger C G 2014 *J. Phys.: Conf. Ser.* **548** 012040

[2] Parigger C G, Lewis J W L, Plemmons D P and Hornkohl J O 1996 Nitric oxide optical breakdown spectra and analysis by the use of the program NEQAIR *Laser Appl. Chem. Bio. Env. Anal.* **3** 85

[3] Hornkohl J O, Parigger C G and Lewis J W L 1991 *J. Quant. Spectrosc. Radiat. Transfer* **46** 405

[4] Parigger C G, Plemmons D H, Hornkohl J O and Lewis J W L 1994 *J. Quant. Spectrosc. Radiat. Transfer* **52** 707

[5] Parigger C G, Plemmons D H, Hornkohl J O and Lewis J W L 1995 *Appl. Opt.* **34** 3331

[6] Park C 1995 *Nonequilibrium Air Radiation (NEQAIR) Program: User's Manual* NASA TM 86707

[7] Laoux C O 1993 Optical diagnostics and radiative emission of air plasmas *PhD Dissertation* (Department of Mechanical Engineering, Stanford University, CA)

[8] Whiting E E 1995 *Private Communication*

[9] Whiting E E, Park C, Liu Y, Arnold J O and Paterson J A 1996 *NEQAIR96, Nonequilibrium Radiative Transport and Spectra Program: User's Manual* NASA TM 1389

[10] Boulous M I, Fauchais P and Pfender E 1994 *Thermal Plasmas—Fundamentals and Applications* (New York: Plenum)

[11] Drakes J A, Pruitt D W, Howard R P and Hornkohl J O 1997 *J. Quant. Spectrosc. Radiat. Transfer* **57** 23

[12] Lewis J W L, Parigger C G, Hornkohl J O and Guan G, 1999 Laser-induced optical breakdown plasma spectra and analyses with the program NEQAIR, *37th Aerospace Sciences Meeting and Exhibit*, AIAA-99-0723

[13] Plemmons D H, Parigger C G, Lewis J W L and Hornkohl J O 1998 *Appl. Opt.* **37** 2493

[14] Ochkin V N 2009 *Spectroscopy of Low Temperature Plasma* (Weinheim: Wiley)

[15] Nemes L, Hornkohl J O and Parigger C G 2005 *Appl. Opt.* **44** 3686

[16] Parigger C G, Woods A C, Witte M J, Swafford L D and Surmick D M 2014 *J. Vis. Exp.* **84** 51250

[17] Nelder J A and Mead R 1965 *Comput. J.* **7** 308

[18] Lagarias J C, Reeds J A, Wright M H and Wright P E 1998 *SIAM J. Optim.* **9** 112

[19] Levin D A, Laux C O and Kruger C H 1999 *J. Quant. Spectrosc. Radiat. Transfer* **61** 377

[20] Hammack S D, Carter C D, Gord J R and Lee T 2012 *Appl. Opt.* **51** 8817

[21] Schultz C, Sick V, Heinze J and Stricker W 1997 *Appl. Opt.* **36** 3227

[22] Battles B E and Hanson R K 1995 *J. Quant. Spectrosc. Radiat. Transfer* **54** 521

[23] Harrington J E, Noble A R, Smith G P, Jeffries J B and Crosley D R 1995 *Evidence for a New No Production Mechanism in Flames* Paper # WSS/CI 95F-193

[24] Rumminger M, Heberle N H, Dibble R W, Smith G P, Jeffries J B and Crosley D R 1995 *Gas Temperature Above a Porous Radiant Burner: Comparison of Measurements and Model Predictions* paper # WSS/CI 95F-212

[25] Labracherie L, Billiotte M and Houas L 1995 *J. Quant. Spectrosc. Radiat. Transfer* **54** 573

Part III

Appendices

Christian G Parigger and James O Hornkohl

Appendix A

Review of angular momentum commutators

The customary starting point for the quantum theory of angular momentum is the commutation formula for the Cartesian components of the angular momentum operator $\mathbf{J}$,

$$\mathbf{J}_i\mathbf{J}_j - \mathbf{J}_j\mathbf{J}_i = i\epsilon_{ijk}\mathbf{J}_k \tag{A.1}$$

where

$$\epsilon_{ijk} = \begin{cases} +1 & i, j, k \text{ in cyclic order} \\ -1 & i, j, k \text{ not in cyclic order.} \\ 0 & \text{if any indices equal} \end{cases}$$

The above commutator property usually defines the angular momentum operator. Coordinate transformations leave the angular momentum operator definition invariant. The conservation law for angular momentum is fundamental. The definition of angular momentum, equation (A.1), is, of course, invariant under specifically spatial translations and rotations. Furthermore, equation (A.1) is invariant under coordinate inversion and time reversal.

Van Vleck's reversed angular momentum method starts with equation (A.1) but then utilizes change of sign of i for angular momentum when a transformation of coordinates to a system attached to a rotating molecule is made,

$$\mathbf{J}_{i'}\mathbf{J}_{j'} - \mathbf{J}_{j'}\mathbf{J}_{i'} = -i\epsilon_{i'j'k'}\mathbf{J}_{k'}. \tag{A.2}$$

Here, the primed index denotes a rotated coordinate. This equation containing the reversed sign of i is known as Klein's [1] anomalous commutation formula.

Two approaches are debated in this book, namely an operator and an algebraic approach [2], without utilizing Klein's anomalous commutation formula. Each approach begins with the standard commutator formula, equation (A.1). The operator approach is included in the text, the algebraic approach is discussed in detail in this appendix. In principle, equation (A.2) can be utilized in building angular momentum theory. However, in analogy to distinction between right- and left-handed coordinate systems, different signs occur. We only use the *standard* sign as indicated in equation (A.1) in computation of a molecular diatomic spectrum [3], i.e. without resorting to use of Klein's anomaly and Van Vleck's reversed angular momentum method.

Here, the reversal of the sign in equation (A.1) is briefly investigated for a unitary and an anti-unitary transformation. The Euler rotation matrix is a real, unitary matrix (see equation (A.4)). The determinate of the Euler rotation matrix is +1 meaning that the sign of vectors is preserved under rotations. A spatial rotation of coordinates is a proper transformation. Conversely, the inversion or parity operator constitutes an improper rotation—this transformation cannot be described exclusively in terms of the Euler angles. However, angular momentum is a pseudo or axial vector, preserving the sign of **J** under improper rotations. The parity operator is also unitary and equation (A.1) is preserved by the parity operator. Time reversal (time inversion or reversal of motion) changes the sign of **J** and it complex-conjugates the imaginary unit due to time reversal being anti-unitary. Thus, equation (A.1) is invariant under time reversal. As shown in texts (e.g. Messiah [4]), the time reversal operator has been designed to be anti-unitary, consequently, preserving the sign of i in commutation formulae.

An algebraic approach reveals that the commutator equation equation (A.1) remains invariant when proper rotation of coordinates is applied. Standard Cartesian coordinates are employed in the particular representation of the rotation matrix, the use of spherical polar coordinates would yield the same results. The laboratory referenced **J** is transformed to the rotated coordinate system by application of the rotation matrix $\mathbf{D}(\alpha\beta\gamma)$,

$$\mathbf{J}' = \mathbf{D}(\alpha\beta\gamma)\mathbf{J}, \tag{A.3}$$

where α, β, and γ are the Euler angles and $\mathbf{D}(\alpha\beta\gamma)$ is an orthogonal matrix whose determinant is +1, Goldstein [5],

$$\mathbf{D}(\alpha\beta\gamma) = \begin{pmatrix} \cos\alpha\cos\beta\cos\gamma - \sin\alpha\sin\gamma & \sin\alpha\cos\beta\cos\gamma + \cos\alpha\sin\gamma & -\sin\beta\cos\gamma \\ -\cos\alpha\cos\beta\sin\gamma - \sin\alpha\cos\gamma & -\sin\alpha\cos\beta\sin\gamma + \cos\alpha\cos\gamma & \sin\beta\sin\gamma \\ \cos\alpha\sin\beta & \sin\alpha\sin\beta & \cos\beta \end{pmatrix}. \tag{A.4}$$

The Euler angles and the matrix $\mathbf{D}(\alpha\beta\gamma)$ used here are those normally used in quantum mechanics, Messiah [4], Davydov [6], Goldstein [5], Rose [7], Brink and Satchler [8], Tinkham [9], Gottfried [10], Baym [11], and Shore and Menzel [12]. This same set of Euler angles is also used by some authors of books on the theory of diatomic spectra, Judd [13], Mizushima [14]. Evaluation of the angular momentum commutation formulae in the rotated system of coordinates gives

$$\mathbf{J}_{i'}\mathbf{J}_{j'} - \mathbf{J}_{j'}\mathbf{J}_{i'} = i\epsilon_{i'j'k'}\mathbf{J}_{k'}. \tag{A.5}$$

This result is obtained from equations (A.1), (A.3), and (A.4). The calculation is simplified somewhat if one notes that for an orthogonal matrix the cofactors, i.e. signed minor determinants, are equal to the corresponding matrix elements of $\mathbf{D}(\alpha\beta\gamma)$ labeled m_{ij}, i.e. m_{ij} is its own cofactor. For example,

$$\begin{aligned} J_{x'}J_{y'} - J_{y'}J_{x'} = i[&(m_{12}m_{23} - m_{13}m_{22})(J_yJ_z - J_zJ_y) \\ &+ (m_{13}m_{21} - m_{11}m_{23})(J_xJ_z - J_zJ_x) \\ &+ (m_{11}m_{22} - m_{12}m_{21})(J_xJ_y - J_yJ_x)], \end{aligned} \tag{A.6}$$

and since

$$\begin{aligned} m_{12}m_{23} - m_{13}m_{22} &= m_{31}, \\ m_{13}m_{21} - m_{11}m_{23} &= -m_{32}, \\ m_{11}m_{22} - m_{12}m_{21} &= m_{33}, \end{aligned} \tag{A.7}$$

the right side of the equation (A.5) reduces to $iJ_{z'}$.

The review of Klein's [1] anomalous formula concludes that reversal of sign of the sign is not required in diatomic molecular spectroscopy. The anomalous sign in the angular momentum commutators does not reveal a novel aspect of the nature of diatomic molecules. Noteworthy is that the anomalous commutation formula remains today a time honored tradition in the theory of molecular spectra as evidenced in several references [10–22]. Klein's anomalous commutators are means by which matrix elements of various operators in the molecular Hamiltonian are obtained, in particular those expressed in terms of angular momentum raising and lowering operators.

References

[1] Klein O 1929 Zur Frage der Quantelung des asymmetrischen Kreisels *Z. Physik* **58** 730

[2] Parigger C G and Hornkohl J O 2010 *Int. Rev. At. Mol. Phys.* **1** 25

[3] Hornkohl J O and Parigger C G 1996 *Am. J. Phys.* **64** 623

[4] Messiah A 1964 *Quantum Mechanics* (Amsterdam: North Holland)

[5] Goldstein H, Poole C P and Safko J L 2001 *Classical Mechanics* 3rd edn (Reading, MA: Addison-Wesley)

[6] Davydov A S 1965 *Quantum Mechanics* (Oxford: Pergamon)

[7] Rose M E 1995 *Elementary Theory of Angular Momentum* (Mineola, NY: Dover)

[8] Brink D M and Satchler G R 1968 *Angular Momentum* (Oxford: Oxford University)

[9] Tinkham M 1964 *Group Theory and Quantum Mechanics* (New York: McGraw Hill)

[10] Gottfried K 1989 *Quantum Mechanics* (Reading, MA: Addison-Wesley)

[11] Baym G 1969 *Lectures on Quantum Mechanics* (London: Benjamin/Cummings)

[12] Shore B W and Menzel D H 1968 *Principles of Atomic Spectra* (Reading, MA: Addison-Wesley)

[13] Judd B 1975 *Angular Momentum Theory for Diatomic Molecules* (New York: Academic)

[14] Mizushima M 1975 *The Theory of Rotating Diatomic Molecules* (New York: Wiley)
[15] Van Vleck J H 1951 The coupling of angular momentum vectors in molecules *Rev. Mod. Phys.* **23** 213
[16] Freed K F 1966 *J. Chem. Phys.* **45** 4214
[17] Kovács I 1969 *Rotational Structure in the Spectra of Diatomic Molecules* (New York: Elsevier)
[18] Hougen J T 2001 *The Calculation of Rotational Energy Levels and Rotational Line Intensities in Diatomic Molecules* National Institute of Standards and Technology http://physics.nist.gov/DiatomicCalculations Originally published as *The Calculation of Rotational Energy Levels and Rotational Line Intensities in Diatomic Molecules*, J T Hougen NBS Monograph 115 (1970)
[19] Carrington A, Levy D H and Miller T A 1970 *Adv. Chem. Phys.* **18** 149
[20] Brown J M and Carrington A 2003 *Rotational Spectroscopy of Diatomic Molecules* (Cambridge: Cambridge University Press)
[21] Lefebvre-Brion H and Field R W 2004 *The Spectra and Dynamics of Diatomic Molecules* (New York: Elsevier/Academic)
[22] Bunker P R and Jensen P 1998 *Molecular Symmetry and Spectroscopy* 2nd edn (Ottawa: NRC)

IOP Publishing

Christian G Parigger and James O Hornkohl

Appendix B

Effects of raising and lowering operators

This appendix addresses details of the effects of raising and lowering operators on standard states $|JM\rangle$ and on elements of the rotation matrix $D^{J^*}_{M\Omega}(\boldsymbol{\alpha}\beta\gamma)$ [1]. The angular momentum raising and lowering operators have the following effects on the standard $|JM\rangle$ states:

$$J_{\pm}|JM\rangle = C_{\pm}(J,\, M)\,|J,\, M \pm 1\rangle, \tag{B.1}$$

where

$$\begin{aligned} C_{\pm}(J,\, M) &= \sqrt{J(J+1) - M(M \pm 1)} \\ &= \sqrt{(J \mp M)(J \pm M - 1)}. \end{aligned} \tag{B.2}$$

This general equation is of course applicable to the diatomic molecule. However, as a result of approximation, one deals with approximate diatomic eigenfunctions. Contained in Van Vleck's method is his discovery that the above standard results are not directly applicable to approximate diatomic eigenfunctions. Typically two magnetic quantum numbers occur for approximate diatomic eigenfunctions, M and Ω in Hund's case (a) or M_N and Λ in case (b).

In modern notation, approximate diatomic angular momentum states are represented by elements of the rotation matrix, $D^{J^*}_{M\Omega}(\alpha\beta\gamma)$, which carry two magnetic quantum numbers, one more than allowed by the nature of angular momentum. Only $\mathbf{J}^2$ and one of its components, by usual convention J_z, commute with the Hamiltonian. It will be important in our approach to find the effects of the raising and lowering operators on elements of the rotation matrix while applying standard theory.

The rotated raising operator, J'_+,

$$J'_+ = J_{x'} + iJ_{y'}, \tag{B.3}$$

doi:10.1088/978-0-7503-1989-8ch22

lowers the Ω quantum number on Hund's case (a) states. See, for example, Van Vleck [2], Judd [3], Mizushima [4], Freed [5], Kovacs [6], Hougen [7], Carrington *et al* [8], Zare *et al* [9], Brown and Howard [10], and Lefebvre-Brion and Field [11]. Similarly, the rotated raising operator N'_+ lowers the Λ quantum number on case (b) kets. Agreement between eigenvalues of Hamiltonian matrices built using these results and experimentally measured term values has firmly established their correctness. Klein's anomalous commutators are often referenced in debating the reason why J'_+ lowers Ω and why N'_+ lowers Λ. However, of interest will be the following equation:

$$J'_\pm \quad D^{J^*}_{M\Omega}(\alpha\beta\gamma) = -C_\mp(J, \Omega) \quad D^{J^*}_{M,\Omega\mp1}(\alpha\beta\gamma), \tag{B.4}$$

which will be derived below. Note that J'_+ lowers Ω, that J'_- raises Ω, and that an unexpected minus sign occurs.

The nature of angular momentum does not allow M and Ω both to be rigorously *good* quantum numbers. This is equivalent to stating that J_z and $J_{z'}$ do not commute. According to definition of angular momentum, J_z does not commute with J_x or J_y, but it is, perhaps, not obvious that J_z and $J_{z'}$ fail to commute. One can easily show, Gottfried [12], that

$$[J_z, J_{z'}] = i\sin(\beta)J_\beta \tag{B.5}$$

$$= i\sin(\beta)[-\sin(\alpha)J_x + \cos(\alpha)J_y]. \tag{B.6}$$

where $J_\beta = \partial/\partial\beta$ is the angular momentum operator for rotation about the first intermediate y-axis. In general, J_z and $J_{z'}$ do not commute. Thus, $D^{J^*}_{M,\Omega}(\alpha\beta\gamma)$ cannot represent a state of angular momentum of a molecule or any other system. The rotation matrix connects two different states of angular momentum.

B.1 Angular momentum operators

Angular momentum operator representations in terms of Euler angles are elaborated. A rotation provides a particularly simple way of expressing a component of angular momentum. The three Euler rotations give the following three components:

$$J_\alpha = -i\frac{\partial}{\partial\alpha} = J_z, \tag{B.7}$$

$$J_\beta = -i\frac{\partial}{\partial\beta}, \tag{B.8}$$

$$J_\gamma = -i\frac{\partial}{\partial\gamma} = J_{z'}. \tag{B.9}$$

Each of these operators is referenced to a different coordinate system. That is, $J_\alpha = J_z$ in the laboratory system, $J_\beta = J_{y_1}$ in the first intermediate system, and $J_\gamma = J_{z'}$ in the fully rotated system. Applying the first Euler rotation to the vector operator $\mathbf{J}$ results in

$$\begin{pmatrix} J_{x_1} \\ J_{y_1} \\ J_{z_1} \end{pmatrix} = \begin{pmatrix} \cos\alpha & \sin\alpha & 0 \\ -\sin\alpha & \cos\alpha & 0 \\ 0 & 0 & 1 \end{pmatrix} \begin{pmatrix} J_x \\ J_y \\ J_z \end{pmatrix}. \tag{B.10}$$

From this one finds

$$J_{y_1} = J_\beta = -\sin\alpha\, J_x + \cos\alpha\, J_y, \tag{B.11}$$

giving J_β in terms of the laboratory coordinates of **J**. Similarly, using the full rotation matrix, equation (A.4), one can express J_γ in laboratory coordinate system,

$$\mathbf{J}' = D(\alpha\beta\gamma)\mathbf{J}, \tag{B.12}$$

$$J_{z'} = J_\gamma = \cos\alpha\sin\beta\, J_x + \sin\alpha\sin\beta\, J_y + \cos\beta\, J_\alpha, \tag{B.13}$$

where the substitution $J_z = J_\alpha$ has been made. Equations (B.11) and (B.13) can be inverted for J_x and J_y:

$$J_x = -i\left(-\cos\alpha\cot\beta\frac{\partial}{\partial\alpha} - \sin\alpha\frac{\partial}{\partial\beta} + \frac{\cos\alpha}{\sin\beta}\frac{\partial}{\partial\gamma}\right), \tag{B.14}$$

$$J_y = -i\left(-\sin\alpha\cot\beta\frac{\partial}{\partial\alpha} + \cos\alpha\frac{\partial}{\partial\beta} + \frac{\sin\alpha}{\sin\beta}\frac{\partial}{\partial\gamma}\right), \tag{B.15}$$

$$J_z = -i\frac{\partial}{\partial\gamma}. \tag{B.16}$$

The method in obtaining these results included evaluation of J_β and J_γ in terms of the laboratory components of **J**. Similarly, $J_{x'}$, $J_{y'}$, and $J_{z'}$ can be obtained by expressing J_α and J_β in terms of the components of $\mathbf{J}'$. The inverse of the full rotation matrix is applied to the rotated vector $\mathbf{J}'$,

$$\mathbf{J} = D^{-1}(\alpha\beta\gamma)\mathbf{J}' = D^\dagger(\alpha\beta\gamma)\mathbf{J}', \tag{B.17}$$

to find J_α in terms of the rotated coordinates of **J**,

$$J_z = J_\alpha = -\sin\beta\cos\gamma\, J_{x'} + \sin\beta\sin\gamma\, J_{y'} + \cos\beta\, J_\gamma. \tag{B.18}$$

The Euler β-rotation is taken about the first intermediate y-axis meaning that the first intermediate and second intermediate y-axes coincide. Thus, J_β can be evaluated in fully rotated coordinates by applying the inverse of the γ rotation matrix to $\mathbf{J}'$,

$$\begin{pmatrix} J_{x_2} \\ J_{y_2} \\ J_{z_2} \end{pmatrix} = \begin{pmatrix} \cos\gamma & -\sin\gamma & 0 \\ \sin\gamma & \cos\gamma & 0 \\ 0 & 0 & 1 \end{pmatrix} \begin{pmatrix} J_{x'} \\ J_{y'} \\ J_{z'} \end{pmatrix}. \tag{B.19}$$

$J_{y_2} = J_{y_1} = J_\beta$, therefore, we find

$$J_\beta = \sin\gamma\, J_{x'} + \cos\gamma\, J_{y'}. \tag{B.20}$$

The two equations in two unknowns are inverted as before. We find for $J_{x'}$ and $J_{y'}$

$$J_{x'} = -i\left(\cos\gamma\cot\beta\frac{\partial}{\partial\gamma} + \sin\gamma\frac{\partial}{\partial\beta} - \frac{\cos\gamma}{\sin\beta}\frac{\partial}{\partial\alpha}\right), \tag{B.21}$$

$$J_{y'} = -i\left(-\sin\gamma\cot\beta\frac{\partial}{\partial\gamma} + \cos\gamma\frac{\partial}{\partial\beta} + \frac{\sin\gamma}{\sin\beta}\frac{\partial}{\partial\alpha}\right), \tag{B.22}$$

$$J_{z'} = -i\frac{\partial}{\partial\gamma}. \tag{B.23}$$

The raising and lowering operators are then constructed using the results above,

$$J_+ = -ie^{i\alpha}\left(-\cot\beta\frac{\partial}{\partial\alpha} + i\frac{\partial}{\partial\beta} + \frac{1}{\sin\beta}\frac{\partial}{\partial\gamma}\right), \tag{B.24}$$

$$J_- = -ie^{-i\alpha}\left(-\cot\beta\frac{\partial}{\partial\alpha} - i\frac{\partial}{\partial\beta} + \frac{1}{\sin\beta}\frac{\partial}{\partial\gamma}\right), \tag{B.25}$$

$$J_{+'} = -ie^{-i\gamma}\left(\cot\beta\frac{\partial}{\partial\gamma} + i\frac{\partial}{\partial\beta} - \frac{1}{\sin\beta}\frac{\partial}{\partial\alpha}\right), \tag{B.26}$$

$$J_{-'} = -ie^{i\gamma}\left(\cot\beta\frac{\partial}{\partial\gamma} - i\frac{\partial}{\partial\beta} - \frac{1}{\sin\beta}\frac{\partial}{\partial\alpha}\right). \tag{B.27}$$

These general results also apply to systems composed of any number of particles. A modification or better *simplification* is required for a system consisting of a single particle (or two particles, since the two-body reduction can always be applied to a system of two particles). The third Euler angle, γ, is superfluous for a single particle, i.e. $\partial/\partial\gamma = 0$. Choosing the first Euler angle to be azimuthal angle ϕ, and the second Euler angle to be the polar angle θ, then equations (B.14)–(B.16) reduce to the familiar textbook equations for the angular momentum operators of a single particle.

Comparison of equations (B.14)–(B.16) with equations (B.21)–(D.10) shows the components of angular momentum are changed by a coordinate transformation. However, the defining commutators for angular momentum in terms of its Cartesian components remain invariant, although the individual components differ. We note that equations (B.7), (B.24), and (B.25) agree with Judd's [3] equation (1.22), but (B.9), (B.26), (B.27) differ in sign from Judd's equation (1.23) presumingly due to the use of the anomalous commutator formula. It appears that Judd obtained his rotated operators in a manner which guaranteed they would obey Klein's anomalous commutation formula.

B.2 Angular momentum commutators and rotation matrix elements

General relations for commutators are typically obtained by applying the commutator to an abstract ket describing a physical state. Alternatively, in the Schrödinger representation, the commutators are obtained by applying differential operators to physical eigenfunctions. This appendix demonstrates how two anomalous results may occur when applying $[J_{x'}, J_{y'}]$ and $[J_y, J_{y'}]$ to the rotation matrix elements $D^{J^*}_{M\Omega}(\alpha\beta\gamma)$.

The commutator $[J_{x'}, J_{y'}]$ is evaluated using

$$[J_{x'}, J_{y'}] = \left[\frac{1}{2}(J'_+ + J'_-), \frac{1}{2i}(J'_+ - J'_-)\right] = -\frac{i}{2}\left[J'_-, J'_+\right], \tag{B.28}$$

and equation (B.4) which for convenience is repeated here

$$J'_\pm\, D^{J^*}_{M\Omega}(\alpha\beta\gamma) = -C_\mp(J, \Omega)\, D^{J^*}_{M,\Omega\mp1}(\alpha\beta\gamma). \tag{B.29}$$

Successive application of the operators in the rotated frame of reference yields the intermediate result

$$\begin{aligned} -\frac{i}{2}\left[J'_-, J'_+\right]D^{J^*}_{M\Omega}(\alpha\beta\gamma) = -\frac{i}{2}(C_+(J, \Omega - 1)C_-(J, \Omega) \\ - C_-(J, \Omega + 1)C_+(J, \Omega))D^{J^*}_{M\Omega}(\alpha\beta\gamma), \end{aligned} \tag{B.30}$$

which after inserting [compare equation (B.2)]

$$C_\pm(J, \Omega) = \sqrt{(J \mp \Omega)(J \pm \Omega + 1)}, \tag{B.31}$$

leads to

$$[J_{x'}, J_{y'}]D^{J^*}_{M\Omega}(\alpha\beta\gamma) = -i\Omega D^{J^*}_{M\Omega}(\alpha\beta\gamma). \tag{B.32}$$

It might be tempting to conclude anomalous commutator relations in the rotated molecular frame from this identity. However, the rotation matrix elements contain two quantum numbers M and Ω, one too many to represent a physical eigenfunction.

Similarly, $[J_y, J_{y'}]$ is evaluated using

$$[J_y, J_{y'}] = \left(-\frac{1}{4}\left[J_+, J'_+\right] - \frac{1}{4}\left[J_-, J'_-\right] + \frac{1}{4}\left[J_+, J'_-\right] + \frac{1}{4}\left[J_-, J'_+\right]\right), \tag{B.33}$$

and applying it to the rotation matrix elements $D^{J^*}_{M\Omega}(\alpha\beta\gamma)$. Note $J_{y'}$ acts on Ω [see equation (B.29)] while J_y acts on M,

$$J_\pm\, D^{J^*}_{M\Omega}(\alpha\beta\gamma) = C_\pm(J, M)\, D^{J^*}_{M\pm1,\,\Omega}(\alpha\beta\gamma). \tag{B.34}$$

The intermediate step is given here

$$
\begin{aligned}
[J_{y}, J_{y'}]D^{J^*}_{M\Omega}(\alpha\beta\gamma) = &+\frac{1}{4}C_{-}(J, \Omega)J_{+}D^{J^*}_{M\ \Omega-1}(\alpha\beta\gamma)\\
&+\frac{1}{4}C_{+}(J, M)J'_{+}D^{J^*}_{M+1\ \Omega}(\alpha\beta\gamma)\\
&+\frac{1}{4}C_{+}(J, \Omega)J_{-}D^{J^*}_{M\ \Omega+1}(\alpha\beta\gamma)\\
&+\frac{1}{4}C_{-}(J, M)J'_{-}D^{J^*}_{M-1\ \Omega}(\alpha\beta\gamma)\\
&-\frac{1}{4}C_{+}(J, \Omega)J_{+}D^{J^*}_{M\ \Omega+1}(\alpha\beta\gamma)\\
&-\frac{1}{4}C_{+}(J, M)J'_{-}D^{J^*}_{M+1\ \Omega}(\alpha\beta\gamma)\\
&-\frac{1}{4}C_{-}(J, \Omega)J_{-}D^{J^*}_{M\ \Omega-1}(\alpha\beta\gamma)\\
&-\frac{1}{4}C_{-}(J, M)J'_{+}D^{J^*}_{M-1\ \Omega}(\alpha\beta\gamma),
\end{aligned}
\qquad (B.35)
$$

which reduces to

$$[J_{y}, J_{y'}]D^{J^*}_{M\Omega}(\alpha\beta\gamma) = 0. \qquad (B.36)$$

Again, one might be tempted to infer from equation (B.36) a general commutator relation that also applies to angular momentum states.

References

[1] Parigger C G and Hornkohl J O 2010 *Int. Rev. At. Mol. Phys.* **1** 25

[2] Van Vleck J H 1951 The coupling of angular momentum vectors in molecules *Rev. Mod. Phys.* **23** 213

[3] Judd B 1975 *Angular Momentum Theory for Diatomic Molecules* (New York: Academic)

[4] Mizushima M 1975 *The Theory of Rotating Diatomic Molecules* (New York: Wiley)

[5] Freed K F 1966 *J. Chem. Phys.* **45** 4214

[6] Kovács I 1969 *Rotational Structure in the Spectra of Diatomic Molecules* (New York: Elsevier)

[7] Hougen J T 2001 *The Calculation of Rotational Energy Levels and Rotational Line Intensities in Diatomic Molecules* National Institute of Standards and Technology http://physics.nist.gov/DiatomicCalculations Originally published as *The Calculation of Rotational Energy Levels and Rotational Line Intensities in Diatomic Molecules,* J T Hougen NBS Monograph 115 (1970)

[8] Carrington A, Levy D H and Miller T A 1970 *Adv. Chem. Phys.* **18** 149

[9] Zare R N, Schmeltekopf A L, Harrop W J and Albritton D L 1973 *J. Mol. Spectrosc.* **46** 37

[10] Brown J M and Howard B J 1976 *Mol. Phys.* **31** 1517

[11] Lefebvre-Brion H and Field R W 2004 *The Spectra and Dynamics of Diatomic Molecules* (New York: Elsevier/Academic)

[12] Gottfried K 1989 *Quantum Mechanics* (Reading, MA: Addison-Wesley)

Appendix C

Modified Boltzmann plots

This appendix elaborates on the use of Boltzmann plots [1] for diatomic molecular spectroscopy inferences of temperature from measured emission spectra. The Boltzmann plot method evaluates temperature by finding the slope from a semi-log graph of recorded spectral lines, e.g. the integrated line shape in determination of hydrogen excitation temperature in atomic spectroscopy. In molecular spectroscopy, usually there are many lines within a spectral resolution of the order of 0.1 nm, typical for laser-plasma emission spectroscopy that uses resolving powers of the order of 5000. Consequently, a modified Boltzmann plot approach [2] is developed for analysis of measured molecular spectra.

C.1 Boltzmann plots

The formal approach starts with an equation containing an exponential,

$$f(x) = A\, e^{-\alpha x}, \tag{C.1}$$

one evaluates the logarithm of both sides, and puts result into point-slope form for a straight line,

$$\begin{aligned} \ln[f(x)] &= -\,\alpha x + \ln A \\ y(x) &= mx + b. \end{aligned} \tag{C.2}$$

In spectroscopy, spontaneous emission from a gas adheres to the spectral radiance, or frequently labeled 'intensity', originating from an upper level, u, to a lower level, l,

$$I_{\mathrm{ul}} = \frac{64\pi^4\,(a_0 e)^2\, c\; C_{\mathrm{abs}}\, N_0}{3Q}\; C_{\tilde{\nu}}\, \tilde{\nu}_{\mathrm{ul}}^4\, S_{\mathrm{ul}}\, e^{-hcF_{\mathrm{u}}/(k_{\mathrm{B}}T)}. \tag{C.3}$$

The individual terms are listed below:

- $a_0 e$—Bohr radius × electronic charge.

- c, h, k_B—speed of light, Planck constant, Boltzmann constant.
- C_{abs}—absolute intensity calibration factor and $C_{\tilde{\nu}}$—relative spectral sensitivity calibration factor.
- N_0—number density, molecules/cm^3.
- Q—partition function of gas.
- $\tilde{\nu}$—wavenumber of spectral line produced by $u \rightarrow l$ transition.
- S_{ul}—electric dipole line strength.
- F_u—upper term value (i.e. energy eigenvalue divided by hc).
- T—absolute temperature.

The linearized spontaneous emission equation,

$$\ln\left(\frac{I_{ul}}{C_{\tilde{\nu}}\,\tilde{\nu}^4\,S_{ul}}\right) = -\frac{hc}{k_B T}F_u - \ln\left(\frac{64\pi^4\,(a_0 e)^2\,cC_{abs}\,N_0}{3Q}\right), \qquad \text{(C.4)}$$

$$y = mx + b,$$

is solved to find the slope m. A Boltzmann plot is a graph of $\ln(I_{ul}/C_{\tilde{\nu}}\tilde{\nu}^4 S_{ul})$ versus the upper term value F_u. The slope is proportional to $-1/T$, of course provided thermodynamic equilibrium is reached,

$$T = -\frac{hc}{k_B m} \qquad \text{(C.5)}$$

As the points would deviate from straight line, one labels the graph and a non-linear Boltzmann plot that would imply departure from thermal equilibrium. Figure C1 illustrates an example of a Boltzmann plot. Figure C2 displays the spectrum from which the Boltzmann plot is constructed.

A Boltzmann plot can be made using the peak intensity or (better) the intensity integrated under each spectral line, but in either case the spectral lines must be fully resolved. The following shows how to make a Boltzmann plot when the spectral lines are not fully resolved. Table C1 shows an example of a line list.

An emission spectral line is broadened by physical processes in the radiating gas and the properties of the spectrometer used to record the spectral line. A widely used measure of spectral line width is the full-width at half-maximum (FWHM). A spectrum is effectively broken into a finite number of pixels between the minimum λ_{min} and the maximum λ_{max} wavelengths in the spectrum (or the minimum or maximum frequencies ν or wavenumbers $\tilde{\nu}$), and the FWHM can be used as the pixel width.

An emission spectrum is computed by evaluating expressions indicated in equation (C.3). One selects a line shape function (i.e. a Gaussian) and a FWHM and computes the contributions of each line. A snipped of a program code utilizes a Gaussian line shape.

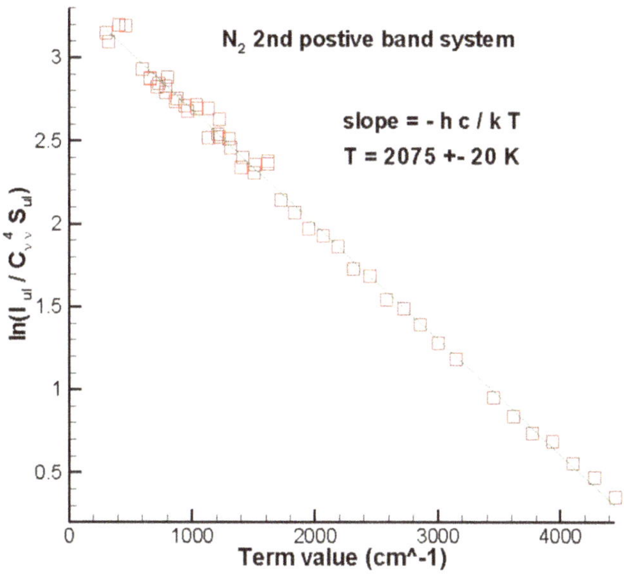

Figure C1. A Boltzmann plot for a spectrum from the N_2 second positive system.

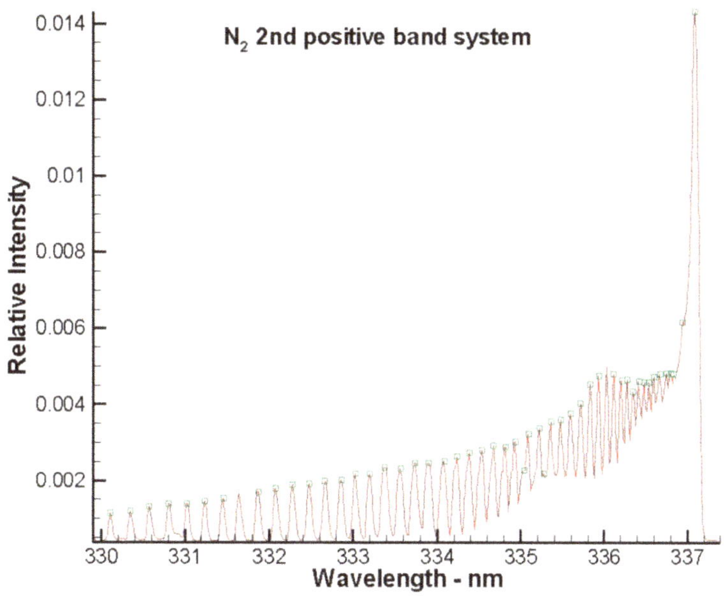

Figure C2. Spectrum for the N_2 second positive system and indicated data points for the Boltzmann plot.

Table C1. Spectral line lists—an example downloaded from NIST.

Wavelength air (Å)	A_{ul} (s^{-1})	E_u (cm^{-1})	E_l (cm^{-1})	g_u
2118.312	1.03×10^7	0	47 192.38	2
2123.362	1.22×10^7	112.061	47 192.38	4
2129.663	1.52×10^7	0	46 940.97	2
2134.733	1.81×10^7	112.061	46 941.55	4
2145.555	2.33×10^7	0	46 593.32	2
2150.699	2.79×10^7	112.061	46 593.95	4
2168.805	3.06×10^7	0	46 093.424	2
2174.028	3.65×10^7	112.061	46 094.312	4
2199.150	1.75×10^6	0	45 457.244	2
2204.590	3.49×10^6	112.061	45 457.244	4
2204.660	4.53×10^7	0	45 344.165	2
2210.046	5.40×10^7	112.061	45 345.594	4
2257.999	3.77×10^6	0	44 273.133	2

```
!  Guassian line shape program segment
   delwn = (wnlo - wnhi) / (npts - 1)
   con = 2.0 * sqrt(alog(2.0))
   do 70 k = 1, kmax
     ndel = nint(2.5*FWHMwn/delwn)
     if (wn(k). lt .wnlo. or .wn(k). gt .wnhi) go to 70
     n0 = nint((wn(k) - wnmin)/delwn) + 1
     nmin = n0 - ndel
     if (nmin. lt .1) nmin = 1
     nmax = n0 + ndel
     if (nmax. gt .npts) nmax = npts
     do n=nmin, nmax
       u = con * (wn(k) - x(n)) / FWHMwn
       if (u. lt .9.21) then
         y(n) = y(n) + peak(k) * exp (-u*u)
         if (y(n). gt .ymax) ymax = y(n)
       end if
     end do
  70 end do
```

C.2 Modified Boltzmann plot

The intensity I_p falling on a pixel (i.e. a small wavenumber range $\Delta\tilde{\nu}$) is the sum of all of the intensities of the spectral lines that contribute to the intensity range $\Delta\tilde{\nu}$,

$$I_p = \sum_{i=1}^{\text{all lines}} \text{const.} f(\tilde{\nu}_i, \Delta\tilde{\nu})\, C_i\, \tilde{\nu}_i^4\, S_i\, e^{-hcF_i/(kT)} \tag{C.6}$$

where $f(\tilde{\nu}_i, \Delta\tilde{\nu})$ is the line shape function. Because the term value F_i varies greatly from line to line, the exponential cannot be taken outside the summation. Therefore, an arbitrary exponential $e^{-hcF_i/(kT)}$ is inserted,

$$I_p = e^{-hcF_j/(kT)} \sum_{i=1}^{\text{all lines}} \text{const.} f(\tilde{\nu}_i, \Delta\tilde{\nu})\, C_i\, \tilde{\nu}_i^4\, S_i\, e^{-hc(F_j-F_i)/(kT)}. \tag{C.7}$$

The exponential $e^{-hcF_j/(kT)}$ can be isolated meaning that for assumed values of the term value F_j and temperature T, one can create a Boltzmann plot. The value of F_j is chosen to be F_i for the spectral line making the biggest contribution to I_p. The value of

$$ln\left(\frac{I_p}{C_i\, \tilde{\nu}^4\, S_i}\right) \Bigg/ \sum_{i=1}^{\text{all lines}} \text{const.} f(\tilde{\nu}_i, \Delta\tilde{\nu}_i)\, e^{-hc(F_j-F_i)/(kT)} \tag{C.8}$$

is plotted versus F_j for a trial T, but there is no reason that the temperature inferred from this plot will equal the trial temperature. However, the new temperature can replace the trial temperature, and the process be repeated. That is, the procedure must be iterated until the trial temperature and the temperature from the Boltzmann plot agree.

An example output from modified Boltzmann plot routine displays the effectiveness of the iteration to find the excitation temperature.

Trial temperature	=	6000.
1 2078.		
2 2073.		
3 2073.		
4 2073.		
5 2073.		
Std. Dev. Boltzmann plot	=	5.1378×10^{-2}
$a0$	=	$3.364 \pm 1.2980 \times 10^{-2}$
Second radiation constant	=	1.438 79
$a1$	=	$-\,6.9404 \times 10^{-4} \pm 6.314 \times 10^{-6}$
Boltzmann plot temperature	=	2073. Std. dev. = ±18.9 K

References

[1] Wiese W L 1991 *Spectrochim. Acta Part B: At. Spectrosc.* **46** 831

[2] Hornkohl J O, Parigger C G and Lewis J W L 1991 *J. Quant. Spectrosc. Radiat. Transfer* **46** 405

Christian G Parigger and James O Hornkohl

Appendix D

Aspects of nitric oxide computations

This appendix communicates a collection of notes on the molecular parameters of NO. These notes contain considerable discussion of the quantum mechanics of the diatomic molecule. The interesting topic of hyperfine structure is ignored.

In the present state of the applied science, the molecular parameters are found by fitting term value differences computed from an approximate Hamiltonian to the best experimental values for the line positions. The computation begins with trial values of the parameters which are then iteratively refined until the differences between the computed and experimental line positions are minimized in the least squares sense.

This appendix discusses results found from fitting some of the best available NO data. The literature on NO is too vast for a comprehensive review. It can only be hoped that the following offers a representative sampling.

D.1 Matrix elements of the Hamiltonian

This section gives a review of the methods of applied quantum mechanics in diatomic spectroscopy. As an example of high quality experimental spectra, reference will be made to the NO rotation–vibration data reported by Amiot, Bacis, and Guelachvili [1]. Their measured line positions are not reported in the journal article, but their data are still available from the Canadian Depository of Unpublished Data. The Macki and Wells [2] data for the (1,0) band are more accurate than the data of Amiot *et al*, but Macki and Wells report only smoothed line positions instead of measured values.

For their analysis of their NO spectra, Amiot, Bacis, and Guelachvili reference the important paper by Zare *et al* [3], which describes the computation of term values by diagonalization of the Hamiltonian. The method of Zare *et al* consists of four steps that are further discussed.

doi:10.1088/978-0-7503-1989-8ch24

In the first step, trial values of the molecular parameters and analytical methods are used to evaluate the matrix elements of the Hamiltonian in Hund's case (a) basis. The second step is an analytical approximation called the Van Vleck transformation which reduces the dimension of the square Hamiltonian matrix to $2(2S + 1) \times 2(2S + 1)$ where S is the total electronic spin quantum number for the dominant state in the basis (if the latter is a Σ state the result of the Van Vleck transformation has the dimensions $(2S + 1) \times (2S + 1)$). In addition to reducing the dimensions of the Hamiltonian, the Van Vleck transformation also splits the Hamiltonian into two independent submatrices of opposite parity.

The ground state of NO is usually modeled as a dominant $^2\Pi$ state which contributes a 4×4 submatrix to the Hamiltonian weakly mixed with a $^2\Sigma^+$ state which contributes a 2×2 submatrix. Thus, in this model the Hund's case (a) representation of the Hamiltonian is a 6×6 matrix. Van Vleck's approximate transformation reduces the Hamiltonian to a 4×4 matrix composed of two independent 2×2 matrices of opposite parity. The third step in the method of Zare *et al* is numerical diagonalization of the two submatrices. The fourth step is a matrix computation which gives corrections to the trial values of the molecular parameters. If the computed corrections are not negligibly small, steps 1–4 are iteratively repeated until the corrections become negligibly small.

The following determinations of NO parameters differ from the method of Zare *et al* in one way. The Van Vleck transformation is skipped. Thus, for the ground state of NO the present computations involve numerical diagonalization of a single 6×6 matrix instead of numerical diagonalization of two 2×2 submatrices.

Various different Hamiltonian models for NO ($X^2\Pi$) will be investigated below but the conclusion is that the accepted model of a dominant $^2\Pi$ state with a small component of a $^2\Sigma^+$ state adequately describes the available experimental observations. Table D1 shows the fit of this model to the NO (1,0) band data of Amiot, Bacis, and Guelachvili [1].

Most of the symbols in the following tables have their standard meanings. $T_v = T_e + G_v$ where T_e is the electronic term value and G_v is the vibrational term value. $\Delta\tilde{\nu}$ is the search tolerance used by the fitting program when it searches for an experimental spectral line to match a predicted line.

An experimental line for which there is no predicted line having the same J' and J'' whose vacuum wavenumber is within the search tolerance $\Delta\tilde{\nu}$ is eliminated from the fit. The table caption entry '405 lines of 419' indicates that 14 experimental lines were rejected from the fit, and σ is the standard deviation of the predicted line positions (vacuum wave number) with respect to the 405 accepted experimental line positions. Values in parenthesis represent 1 standard deviation expressed in the last digits. Values in square brackets were held fixed during the fitting process. Other symbols in the table will be defined below. The ratio of the standard deviation of the fit to the band origin, $\sigma/\tilde{\nu}_{10} \approx 2 \times 10^{-7}$.

Most of the richness of the typical diatomic spectrum is produced by the rotational Hamiltonian, the term in the diatomic Hamiltonian representing the kinetic energy of rotation of the nuclei, given by

Table D1. The molecular parameters for the $v = 0$ and $v = 1$ states of the ground electronic state of NO obtained by fitting the model of a dominant $^2\Pi$ state weakly mixed with a $^2\Sigma^+$ state to the (1,0) band data of Amiot, Bacis, and Guelachvili [1]. The parameters for the $^2\Sigma^+$ state were held fixed to those values found from a fit to the $A^2\Sigma^+(v = 0) \leftrightarrow X^2\Pi(v = 0)$ γ-system data of Engleman *et al* [4] The upper state extends to $0.5 \leqslant J' \leqslant 40.5$, includes 405 of 429 lines, with an accuracy of $\Delta\tilde{\nu} = 0.001$ cm^{-1} and $\sigma = 0.00038$ cm^{-1}. The term value difference equals $\tilde{\nu}_{10} = 1875.9731$.

$X^2\Pi(v = 1)$	$T_v = 2824.6331(5)$
$B_v = 1.678\,571\,5(27)$	$D_v = 5.4893(26) \times 10^{-6}$
$A_v = 122.7026(21)$	$A_{vJ} = 3.475(38) \times 10^{-4}$
$\langle AL \rangle = -177.8(1.0)$	$\langle AL \rangle_J = 4.2(1.8) \times 10^{-3}$
$\langle BL \rangle = -1.3922(76)$	$\langle BL \rangle_J = -4.2(1.3) \times 10^{-5}$
$X^2\Pi(v = 0)$	$[T_v = 948.66]$
$B_v = 1.696\,144\,5(30)$	$D_v = 5.4735(29) \times 10^{-6}$
$A_v = 122.9495(23)$	$A_{vJ} = 3.565(42) \times 10^{-4}$
$\langle AL \rangle = -180.4(1.1)$	$\langle AL \rangle_J = 3.4(2.0) \times 10^{-3}$
$\langle BL \rangle = -1.433(84)$	$\langle BL \rangle_J = -3.1(1.5) \times 10^{-5}$
$A^2\Sigma^+(v = 0)$	$[T_v = 45087.65]$
$[B_v = 1.986\,312]$	$[D_v = 5.575 \times 10^{-6}]$
$[\gamma_v = -1.34 \times 10^{-4}]$	

$$H_{\text{rot}} = B(r)\,\mathbf{R}^2 \tag{D.1}$$

in which $\mathbf{R}$ is the orbital angular momentum operator for the nuclei whose motion has been reduced to that of a single, fictitious particle of reduced mass μ,

$$\mu = \frac{m_a m_b}{m_a + m_b} \tag{D.2}$$

of the nuclei whose masses are m_a and m_b, and

$$B(r) = \frac{\hbar}{4\pi c\mu r^2} \tag{D.3}$$

where r is the internuclear distance. The rotational parameter $B(r)$ has the units of energy, but the right side of the above equation has been divided by hc (Planck's constant $\times$ the speed of light) to give $B(r)$ the spectroscopist's unit of energy which has the units of reciprocal length.

When, as has been assumed here, the normally very small influence of nuclear spin on the term values can safely be ignored, the total angular momentum,

$$\mathbf{J} = \mathbf{L} + \mathbf{R} + \mathbf{S} \tag{D.4}$$

is the sum of the total electronic orbital angular momentum $\mathbf{L}$, the total nuclear orbital angular momentum $\mathbf{R}$, and the total electronic spin $\mathbf{S}$. The rotational Hamiltonian becomes [5]

$$\begin{aligned} H_{\mathrm{rot}} &= B(r)(\mathbf{J} - \mathbf{L} - \mathbf{S})^2 \\ &= B(r)\Big[\mathbf{J}^2 + \frac{L_+L_- + L_-L_+}{2} + L_z^2 + \mathbf{S}^2 \\ &\quad - (J_+L_- + J_-L_+ + 2J_zL_z) \\ &\quad - (J_+S_- + J_-S_+ + 2J_zS_z) + (L_+S_- + L_-S_+ + 2L_zS_z)\Big] \end{aligned} \tag{D.5}$$

$$\begin{aligned} &= B(r)\Big[\mathbf{J}'^2 + \frac{L'_+L'_- + L'_-L'_+}{2} + L_{z'}^2 + \mathbf{S}'^2 - \left(J'_+L'_- + J'_-L'_+ + 2J_{z'}L_{z'}\right) \\ &\quad - \left(J'_+S_-' + J'_-S_+' + 2J_{z'}S_{z'}\right) + \left(L'_+S_-' + L'_-S_+' + 2L_{z'}S_{z'}\right)\Big]. \end{aligned} \tag{D.6}$$

where unprimed components are in the laboratory coordinate system, primed components are in a coordinate system attached to the molecule, and the raising and lowering operators are defined by, for example,

$$J_\pm = J_x \pm J_y, \tag{D.7}$$

with similar equations for $S_\pm$ and $L_\pm$. The analytical form of each term in the Hamiltonian is preserved under a coordinate transformation (compare equations (D.5) and (D.6)). Matrix elements of Hamiltonian terms are much easier to analytically evaluate in molecular coordinates than in the laboratory coordinates.

As explained in many quantum mechanics texts, specification of the states of a quantum system is normally done with the standard $|nJM\rangle$ states of the system where J is the quantum number for the total angular (the quantity which obeys the conservation law for angular momentum), M is the quantum number for the laboratory z component of $\mathbf{J}$, and n denotes all other required quantum numbers. The exact equation for the standard $|nJM\rangle$ states of the diatomic molecule is [5]

$$\langle \mathbf{r}_1, \mathbf{r}_2, \ldots, \mathbf{r}_{\mathcal{N}}, \mathbf{r}|nJM\rangle = \sum_{\Omega=-J}^{J} \langle \mathbf{R}'_e r|n\rangle\, D_{M\Omega}^{J^*}(\alpha\beta\gamma) \tag{D.8}$$

where $\mathbf{r}_1, \mathbf{r}_2, \ldots, \mathbf{r}_{\mathcal{N}}$ are the laboratory coordinates of the $\mathcal{N}$ electrons, $\mathbf{r}$ is the internuclear vector, α, β, and γ are the Euler angles, R'_e represent $3\mathcal{N} - 1$ rotated electronic coordinates (γ is the missing electronic coordinate), and $D_{M\Omega}^{J}(\alpha\beta\gamma)$ is the matrix element of the rotation operator (note that its complex conjugate appears in the above equation). Hund's case (a) basis is defined by [5]

$$\langle \mathbf{r}_1, \mathbf{r}_2, \ldots, \mathbf{r}_{\mathcal{N}}, \mathbf{r}|nJM\Omega\Lambda S\Sigma\rangle = \sqrt{\frac{2J+1}{8\pi^2}}\langle \mathbf{R}'_e r|n\rangle\, |S\Sigma\rangle\, D_{M\Omega}^{J^*}(\alpha\beta\gamma) \tag{D.9}$$

in which Ω is the quantum number for the z' component of J and Σ is the quantum number for the z' component of S. The relationship

$$J_{z'} = L_{z'} + S_{z'} \tag{D.10}$$

holds because the z' axis lies on the internuclear axis and the nuclei have no internuclear component of orbital angular momentum (i.e. $R_{z'} = 0$). Thus,

$$\Omega = \Lambda + \Sigma \tag{D.11}$$

and Λ, the quantum number for $L_{z'}$ is also a case (a) quantum number although it does not explicitly appear on the right side of equation (D.9).

The Born–Oppenheimer approximation has not been applied in either the exact equation for the $|nJM\rangle$ diatomic states, equation (D.8), or the case (a) basis, equation (D.9). The smallness of the electronic mass in comparison to the nuclear mass produces an approximate separation of the electronic and nuclear motions. In the usual exposition of the Born–Oppenheimer approximation the approximate separation of the electronic and nuclear motions involves all of the electronic and nuclear internal coordinates. However, in the present formulation the Euler angles play no part in the Born–Oppenheimer separation which is concerned exclusively with the eigenfunction $\langle \mathbf{R}'_e r|n\rangle$. The quantum number v, the quantum number associated with the internuclear distance r, is extracted from the collection of quantum numbers represented by the symbol n, and the Born–Oppenheimer approximate separation of $\langle \mathbf{R}'_e r|nv\rangle$ is then made,

$$\begin{aligned} \left\langle \mathbf{R}'_e r|nv\right\rangle &\approx \left\langle \mathbf{R}'_e;\, r|n\right\rangle \langle r|v\rangle \\ &= \psi_n\!\left(\mathbf{R}'_e;\, r\right) \psi_v(r) \end{aligned} \tag{D.12}$$

in which the semicolon is a notational device indicating that the electronic eigenfunction $\psi_e(\mathbf{R}'_e;\, r)$ is a parametric function of the internuclear distance r. This means that the electronic Schrödinger equation has a different solution for each value of the internuclear distance. Because the electronic eigenfunction is a parametric function of r, all electronic matrix elements are functions of r.

Application of the Born–Oppenheimer approximation to the case (a) basis gives

$$\begin{aligned} \langle \mathbf{r}_1, \mathbf{r}_2, \ldots, \mathbf{r}_N, \mathbf{r}|nvJM\Omega\Lambda S\Sigma\rangle = \sqrt{\frac{2J+1}{8\pi^2}} \left\langle \mathbf{R}'_e;\, r|n\right\rangle \\ \langle r|v\rangle |S\Sigma\rangle\, D^{J*}_{M\Omega}(\alpha\beta\gamma). \end{aligned} \tag{D.13}$$

Matrix elements of the rotational Hamiltonian can be calculated using equations (D.6) and (D.13) (table D2). Zare *et al* [3] give tables of case (a) matrix elements. They are also given by Lefebvre-Brion and Field [6]. Sign ambiguities between Zare *et al* and Lefebvre-Brion and Field can be resolved through application of equations (D.6) and (D.9) [5]. Table D3 gives example matrix elements of H_{rot} in a state consisting of a mixture of $^2\Pi$ and $^2\Sigma$ case (a) basis states. The diagonal matrix elements of the rotational Hamiltonian, $\langle nJM\Omega\Lambda S\Sigma|H_{\mathrm{rot}}|nJM\Omega\Lambda S\Sigma\rangle$, are given by

Table D2. Matrix elements of the rotational Hamiltonian, equation (D.6).

$$\begin{aligned}
\langle nvJM\Omega\Lambda S\Sigma|H_{\mathrm{rot}}|nvJM\Omega\Lambda S\Sigma\rangle &= \langle nvJM\Omega\Lambda S\Sigma|B(r) \\
&\quad \times (\mathbf{J}^2 + L_{z'}^2 + \mathbf{S}^2 - 2J_{z'}L_{z'} - 2J_{z'}S_{z'} + 2L_{z'}S_{z'}) \\
&\quad \times |nvJM\Omega\Lambda S\Sigma\rangle \\
&= \langle nv|B(r)|nv\rangle[J(J+1) - \Omega^2 + S(S+1) - \Sigma^2] \\
&= B_v[J(J+1) - \Omega^2 + S(S+1) - \Sigma^2] \\
\langle nvJM\Omega\Lambda S\Sigma|H_{\mathrm{rot}}|nvJM, \Omega \pm 1, \Lambda S, \Sigma \pm 1\rangle &= \langle nvJM\Omega\Lambda S\Sigma| - B(r)(J_+'S_-' + J_-'S_+') \\
&\quad |nvJM, \Omega \pm 1, \Lambda S, \Sigma \pm 1\rangle \\
&= \langle nv|B(r)|nv\rangle\ C_\pm(J\Omega)\ C_\pm(S\Sigma) \\
&= B_v\sqrt{J(J+1) - \Omega(\Omega \pm 1)} \\
&\quad \sqrt{S(S+1) - \Sigma(\Sigma \pm 1)} \\
\langle nvJM\Omega\Lambda S\Sigma|H_{\mathrm{rot}}|n'v'JM, \Omega \pm 1, \Lambda \pm 1, S\Sigma\rangle &= \langle nvJM\Omega\Lambda S\Sigma| - B(r)(J_+'L_-' + J_-'L_+') \\
&\quad |n'v'JM, \Omega \pm 1, \Lambda \pm 1, S\Sigma\rangle \\
&= \langle nv|B(r)(L_+' + L_-')|n'v'\rangle\ C_\pm(J\Omega) \\
&= \langle BL\rangle\sqrt{J(J+1) - \Omega(\Omega \pm 1)} \\
\langle nvJM\Omega\Lambda S\Sigma|H_{\mathrm{rot}}|n'v'JM\Omega, \Lambda \pm 1, S, \Sigma \mp 1\rangle &= \langle nvJM\Omega\Lambda S\Sigma|B(r)(L_+'S_-' + L_-'S_+') \\
&\quad |n'v'JM\Omega, \Lambda \pm 1, S, \Sigma \mp 1\rangle \\
&= \langle nv|B(r)(L_+' + L_-')|n'v'\rangle\ C_\mp(S\Sigma) \\
&= \langle BL\rangle\sqrt{S(S+1) - \Sigma(\Sigma \mp 1)}
\end{aligned}$$

Table D3. Another fit to the data of Amiot, Bacis, and Guelachvili [1] with a different model Hamiltonian. In this example the weak ${}^2\Sigma^+$ contribution comes from a fictitious ${}^2\Sigma^+$ state. Angular momentum range $0.5 \leqslant J' \leqslant 40.5$, includes 405 of 419 lines, accuracy $\Delta\tilde{\nu} = 0.001\ \mathrm{cm}^{-1}$ and $\sigma = 0.000\ 38\ \mathrm{cm}^{-1}$.

$X^2\Pi(v = 1)$	$T_v = 2824.6331(5)$
$B_v = 1.678\ 571\ 5(27)$	$D_v = 5.4893(26) \times 10^{-6}$
$A_v = 122.7029(21)$	$A_{vJ} = 3.471(38) \times 10^{-4}$
$AL_+ = -223.8(1.3)$	$AL_{+J} = 6.9(2.3) \times 10^{-3}$
$BL_+ = -1.7549(95)$	$BL_{+J} = -4.(1.6) \times 10^{-5}$
$X^2\Pi(v = 0)$	$[T_v = 948.66]$
$B_v = 1.696\ 144\ 4(29)$	$D_v = 5.4735(29) \times 10^{-6}$
$A_v = 122.9498(23)$	$A_{vJ} = 3.561(42) \times 10^{-4}$
$AL_+ = -225.4(1.4)$	$AL_{+J} = 5.8(2.6) \times 10^{-3}$
$BL_+ = -1.79(10)$	$BL_{+J} = -3.9(1.8) \times 10^{-5}$
$A^2\Sigma^+(v = 0)$	$[T_v = 70000.0]$
$[B_v = 1.2]$	$[D_v = 5.0 \times 10^{-6}]$

$$B_v \left\langle nJM\Omega\Lambda S\Sigma|\mathbf{J}^2 + L_{z'}^2 - 2J_{z'}L_{z'} - 2J_{z'}S_{z'} + 2L_{z'}S_{z'}|JM\Omega\Lambda S\Sigma\right\rangle$$
$$= B_v\,[J(J+1) - \Omega^2 + S(S+1) - \Sigma^2], \tag{D.14}$$

where the rotational constant B_v is defined by

$$B_v = \frac{\hbar}{4\pi c\mu}\langle nv|r^{-2}|nv\rangle = \frac{\hbar}{4\pi c\mu}\langle v|r^{-2}|v\rangle. \tag{D.15}$$

Because the internuclear distance is held constant for evaluation of an electronic matrix element, the rotational constant is diagonal with respect to the electronic quantum numbers n.

The operations involving **J** and **S** are straightforward. Operation of $\mathbf{J}^2$, J_z, and $J_{z'}$ on the standard $|nJM\rangle$ and $|nJ\Omega\rangle$ states obey,

$$\mathbf{J}^2\,|nJM\rangle = J(J+1)\,|nJM\rangle \tag{D.16}$$

$$J_z\,|nJM\rangle = M\,|nJM\rangle \tag{D.17}$$

$$J_{z'}\,|nJ\Omega\rangle = \Omega\,|nJ\Omega\rangle \tag{D.18}$$

The only possible point of confusion here is the standard state $|nJ\Omega\rangle$ which is referenced to the rotated, molecule fixed, z axis (i.e. the z'-axis) instead of the laboratory z-axis. For any system, the standard $|nJM\rangle$ and $|nJ\Omega\rangle$ states are related by

$$\langle \mathbf{r}_1, \mathbf{r}_2, \ldots, \mathbf{r}_{\mathcal{N}}|nJM\rangle = \sum_{\Omega=-J}^{J} \langle \mathbf{r}_1', \mathbf{r}_2', \ldots, \mathbf{r}_{\mathcal{N}}'|nJ\Omega\rangle\, D_{M\Omega}^{J^*}(\alpha\beta\gamma) \tag{D.19}$$

where, for the moment, $\mathcal{N}$ is the number of particles having spatial coordinates in the system. This is a general result holding for any quantum system (see, e.g. Thompson's [7] equation (6.19) or equation (xx) of [5]).

The operations involving **L** are more complicated. The operators $\mathbf{J}^2$ and $J_\pm$ act on the rotation matrix element but not $\langle \mathbf{R}_e' r|nv\rangle$ or the electronic spin states $|S\Sigma\rangle$. The operator $\mathbf{S}^2$ affects nothing but the spin ket $|S\Sigma\rangle$. The total electronic orbital angular momentum,

$$\mathbf{L} = \sum_{i=1}^{\mathcal{N}} \mathbf{l}_i \tag{D.20}$$

in which l_i is the orbital angular momentum of the ith electron is a function of $3\mathcal{N}$ coordinates where $\mathcal{N}$ is the number of electrons. The symbol $\mathbf{R}_e'$ represents only $3\mathcal{N} - 1$ electronic coordinates, and the third Euler angle γ is the remaining electronic coordinate. Thus, $3\mathcal{N} - 1$ components of L operate on $\langle \mathbf{R}_e'|nv\rangle$ and the remaining component l_γ acts on $D_{M,\Omega}^{J^*}(\alpha\beta\gamma)$. A Clebsch–Gordon expansion of the rotation matrix element gives

$$\langle \mathbf{r}_1, \mathbf{r}_2, \ldots, \mathbf{r}_N, \mathbf{r}|nJM\Omega\Lambda S\Sigma\rangle = \sqrt{\frac{2J+1}{8\pi^2}}\langle \mathbf{R}_e' r|nv\rangle\,|S\Sigma\rangle \times \sum_{M_N=-N}^{N}\sum_{M_S=-S}^{S}\langle NM_N SM_S|JM\rangle \times \langle N\Lambda S\Sigma|J\Omega\rangle\, D_{M_N\Lambda}^{N*}(\alpha\beta\gamma)\, D_{M_S\Sigma}^{S*}(\alpha\beta\gamma). \tag{D.21}$$

in which N, the laboratory referenced M_N, and the molecule referenced Λ are the quantum numbers associated with the total orbital angular momentum $\mathbf{N}$,

$$\mathbf{N} = \mathbf{L} + \mathbf{R} \tag{D.22}$$

For the one electron for which γ is the angular coordinate of rotation about the internuclear axis,

$$\begin{aligned} l_{z'}\, D_{M_N\Lambda}^{N*}(\alpha\beta\gamma) &= -i\frac{\partial}{\partial\gamma} D_{M_N\Lambda}^{N*}(\alpha\beta\gamma) \\ &= -i\frac{\partial}{\partial\gamma} e^{iM_N\alpha}\, d_{M_N\Lambda}^{N}(\beta)\, e^{i\Lambda\gamma} \\ &= \Lambda D_{M_N\Lambda}^{N*}(\alpha\beta\gamma). \end{aligned} \tag{D.23}$$

Thus, operation of l_γ in the case (a) basis is given by

$$l_\gamma\,|nJM\Omega\Lambda S\Sigma\rangle = \Lambda\,|nJM\Omega\Lambda S\Sigma\rangle. \tag{D.24}$$

Calculation of off-diagonal matrix elements of the rotational Hamiltonian is complicated by the non-standard behavior of the case (a) states under operation of the raising and lowering operators. The rotation matrix element carries two magnetic quantum numbers, one more than mathematically admitted by the definition of angular momentum. The familiar behavior of standard $|nJM\rangle$ under raising and lowering operators,

$$\begin{aligned} J_\pm\,|nJM\rangle &= \sqrt{J(J+1) - M(M\pm 1)}\,|nJ, M\pm 1\rangle \\ &= C_\pm(JM)\,|nJ, M\pm 1\rangle \end{aligned} \tag{D.25}$$

does not hold for the rotation matrix element even though it looks like an angular momentum eigenfunction because it is specified in terms of angular coordinates and quantum numbers. The rotation matrix elements obey [5]

$$J_\pm D_{M\Omega}^{J}(\alpha\beta\gamma) = -C_\mp(JM)\, D_{M\mp 1,\Omega}^{J}(\alpha\beta\gamma), \tag{D.26}$$

$$J_\pm' D_{M\Omega}^{J}(\alpha\beta\gamma) = C_\pm(J\Omega)\, D_{M,\Omega\pm 1}^{J}(\alpha\beta\gamma), \tag{D.27}$$

$$J_\pm D_{M\Omega}^{J*}(\alpha\beta\gamma) = C_\pm(JM)\, D_{M\pm 1,\Omega}^{J*}(\alpha\beta\gamma), \tag{D.28}$$

$$J_\pm' D_{M\Omega}^{J*}(\alpha\beta\gamma) = -C_\mp(J\Omega)\, D_{M,\Omega\mp 1}^{J*}(\alpha\beta\gamma). \tag{D.29}$$

These equations show that the rotation matrix element does not have the mathematical properties of an angular momentum eigenfunction. One can conclude, with the support of mathematical certainty, that $D^J_{M\Omega}(\alpha\beta\gamma)$ is not an angular momentum eigenfunction. One armed with the ability to count to two can draw the same conclusion from the symbol $D^J_{M\Omega}(\alpha\beta\gamma)$. Nevertheless, essentially without exception in diatomic literature the rotation matrix element is treated as if it were an angular momentum eigenfunction, and this is the source of considerable confusion.

Using equation (D.29) for the non-standard case (a) basis states but using the standard result

$$S_{\pm}\,|S\Sigma\rangle = C_{\pm}(S\Sigma)\,|S,\,\Sigma \pm 1\rangle \tag{D.30}$$

for the electronic spin states because $|S\Sigma\rangle$ is standard angular momentum state of the type given in equation (D.25), one finds

$$\begin{aligned}\langle nvJM\Omega\Lambda S\Sigma|H_{\text{rot}}|nvJM,\,\Omega \mp 1,\,\Lambda S,\,\Sigma \pm 1\rangle &= -\,\langle nJM\Omega\Lambda S\Sigma|J'_+S_-' \\ &\quad + J'_-S_+'|nJM,\,\Omega \mp 1,\,\Lambda S,\,\Sigma \pm 1\rangle \\ &= B_v\sqrt{J(J+1) - \Omega(\Omega \pm 1)} \\ &\quad \times \sqrt{S(S+1) - \Sigma(\Sigma \pm 1)}.\end{aligned} \tag{D.31}$$

$$\begin{aligned}\langle nvJM\Omega\Lambda S\Sigma|H_{\text{rot}}|n'v'JM\Omega,\,\Lambda \mp +1,\,S,\,\Sigma \pm 1\rangle &= -\,\langle nvJM\Omega\Lambda S\Sigma|L'_+S_-' \\ &\quad + L'_-S_+'|nJM,\,\Omega \mp 1,\,\Lambda S,\,\Sigma \pm 1\rangle \\ &= \langle nv|B(r)(L'_+ + L'_-)|n'v'\rangle \\ &\quad \times \sqrt{S(S+1) - \Sigma(\Sigma - 1)} \\ &\quad + \langle nv|B(r)(L'_+ + L'_-)|n'v'\rangle \\ &\quad \times \sqrt{S(S+1) - \Sigma(\Sigma + 1)}\end{aligned} \tag{D.32}$$

$$= \langle BL\rangle\left[\sqrt{S(S+1) - \Sigma(\Sigma-1)} + \sqrt{S(S+1) - \Sigma(\Sigma+1)}\right] \tag{D.33}$$

where

$$\langle BL\rangle = \langle nv|B(r)(L'_+ + L'_-)|n'v'\rangle \tag{D.34}$$

Equation (D.31) and the example given in table D4 show that Hund's case (a) matrix representation of $\mathbf{R}^2$ is non-diagonal (table D5). Therefore, the matrix of the rotational Hamiltonian, equation (D.1), is also non-diagonal. Strictly speaking, Hund's case (a) basis is not a valid eigenfunction. At best, it is a poor physical approximation. This is no way detracts from its use as a basis. Before widespread availability of digital hardware and numerical algorithms, diagonalization of the 6×6 matrix of table D4 presented an essentially impossible task, but today numerical diagonalization of this matrix is trivial with even a very modest computer. In current practice, the fact that Hund's case (a) basis does not yield a diagonal Hamiltonian matrix adds only the single line of code, `CALL JACOBI`, to a computer program.

Table D4. Matrix elements of $\mathbf{R}^2$ in a state which is a mixture of $^2\Pi$ and $^2\Sigma$ case (a) basis states.

Λ'			-1	-1	1	1	0	0
	Σ'		$-1/2$	$1/2$	$-1/2$	$1/2$	$-1/2$	$1/2$
		Ω'	$-3/2$	$-1/2$	$1/2$	$3/2$	$-1/2$	$1/2$
Λ	Σ	Ω						
-1	$-1/2$	$-3/2$	119.000	10.954	0	0	0	0
-1	$1/2$	$-1/2$	10.954	121.000	0	0	0	0
1	$-1/2$	$1/2$	0	0	121.000	10.954	0	0
1	$1/2$	$3/2$	0	0	10.954	121.000	0	0
0	$-1/2$	$-1/2$	0	0	0	0	121.000	11.000
0	$1/2$	$1/2$	0	0	0	0	11.000	121.000

Table D5. Matrix elements of $\mathbf{R}^2$ in a state which is a mixture of $^3\Pi$ and $^3\Sigma$ case (a) basis states.

Λ'			-1	-1	-1	1	1	1	0	0	0
	Σ'		-1	0	1	-1	0	1	-1	0	1
		Ω'	-2	-1	0	0	1	2	-1	0	1
Λ	Σ	Ω									
-1	-1	-2	107.000	14.697	0	0	0	0	0	0	0
-1	0	-1	14.697	111.000	14.832	0	0	0	0	0	0
-1	1	0	0	14.832	111.000	0	0	0	0	0	0
1	-1	0	0	0	0	111.000	14.832	0	0	0	0
1	0	1	0	0	0	14.832	111.000	14.697	0	0	0
1	1	2	0	0	0	0	14.697	107.000	0	0	0
0	-1	-1	0	0	0	0	0	0	110.000	14.832	0
0	0	0	0	0	0	0	0	0	14.832	112.000	14.832
0	1	1	0	0	0	0	0	0	0	14.832	110.000

Table D3 gives the results of fitting the same data of Amiot *et al* used in table D1 but with a different $^2\Sigma$ state. A comparison between tables D1 and D3 shows that only the mixing parameters AL_+ and BL_+ and the associated corrections for centrifugal stretching AL_{+J} and BL_{+J} are significantly influenced by the change in the parameters of the $^2\Sigma$ state.

Zare *et al* give the equations

$$o_v^{\Sigma} = \sum_{n'v'} \frac{\langle n'\ {}^{2S+1}\Pi\ v'\ J'|\frac{1}{2}AL_+|n\ {}^{2S+1}\Sigma\ v\ J\rangle^2}{E_{nvJ} - E_{n'v'J'}} \tag{D.35}$$

$$p_v^{\Sigma} = 4\sum n'v' \frac{\langle n'\ {}^{2S+1}\Pi\ v'\ J'|\frac{1}{2}AL_+|n^{2S+1}\Sigma v J\rangle\langle n'\ {}^{2S+1}\Pi\ v'\ J'|BL_+|n\ {}^{2S+1}\Sigma\ v\ J\rangle}{E_{nvJ} - E_{n'v'J'}} \tag{D.36}$$

$$q_v^{\Sigma} = 2 \sum_{n'v'} \frac{\langle n' \ {}^{2S+1}\Pi \ v' \ J' | AL_{+} | n \ {}^{2S+1}\Sigma \ v \ J \rangle^2}{E_{nvJ} - E_{n'v'J'}} \tag{D.37}$$

for the Van Vleck transformation.

References

[1] Guelachvili G, Amiot C and Bacis R 1978 *Can. J. Phys.* **56** 251

[2] Macki A G and Wells J S 1991 *Wavenumber Calibration tables from Heterodyne Frequency Measurements* NIST special publication 821, Washington, DC

[3] Zare R N, Schmeltekopf A L, Harrop W J and Albritton D L 1973 *J. Mol. Spectrosc.* **46** 37

[4] Engleman R Jr, Rouse P E, Peek H M and Baiamonte V D 1970 *Eta and Gamma Band Systems of Nitric Oxide* Los Alamos Sci. Lab. Report LA-4364

[5] Hornkohl J O and Parigger C G 1996 *Am. J. Phys.* **64** 623

[6] Lefebvre-Brion H and Field R W 2004 *The Spectra and Dynamics of Diatomic Molecules* (New York: Elsevier/Academic)

[7] Thompson W J 1994 *Angular Momentum* (New York: Wiley)

Christian G Parigger and James O Hornkohl

Appendix E

Parity in diatomic molecules

This appendix communicates the application of the parity operator to the general diatomic eigenfunction [1]. The parity eigenvalue is a product of two factors, one that depends on the total angular momentum quantum number and a second constant factor which can be interpreted as the intrinsic parity of the molecule. These new results allow one to rigorously design an algorithm for the computation of diatomic spectra by utilizing that allowed transitions have non-vanishing rotational line strengths.

E.1 Introduction

The diatomic Hamiltonian matrix is historically parity-partitioned, thereby giving parity a more important role in diatomic spectroscopy than in atomic spectroscopy, e.g. see Zare *et al* [2]. Typically, a diatomic line list will include rotational parity designations for the lower levels [3]. Several authors, for instance Hougen [4], Røeggen [5], Judd [6] and Larsson [7], have presented treatments of diatomic parity using the approximate, Born–Oppenheimer separation of the diatomic eigenfunction into rotational, vibrational and electronic factors.

In this work, operation of the parity operator on the general Wigner–Witmer [8] diatomic eigenfunction is used to yield the parity eigenvalues that are composed of a constant and an angular momentum dependent part. The computation of diatomic molecular spectra is accomplished without the need of explicitly including parity selection rules. The fundamental Wigner–Witmer diatomic eigenfunction simplifies the determination of rotational line strengths, i.e. Hönl–London factors. Allowed transitions are governed by non-zero rotational line strengths.

doi:10.1088/978-0-7503-1989-8ch25

E.2 Parity operator

The discrete parity operation can be accomplished with a rotation and a reflection. The parity operator, $\mathcal{P}$, can be written as a product,

$$\mathcal{P} = C_2\, \sigma_v. \tag{E.1}$$

The determinant of the matrix representations $\sigma_v(y, z)$ and $C_2(x)$ in laboratory xyz-coordinates are -1 and $+1$, respectively. The C_2 operator is a proper rotation that can be expressed as a discrete transformation of Euler angles. The Euler angles are the arguments of the Wigner D-function used to formulate rotational symmetry, consequently, the eigenvalues of C_2 are controlled by the angular momentum, J. The $\sigma_v(y, z)$ operation results in a constant factor, and the C_2 operation yields the angular momentum dependent part of the parity eigenvalue.

E.3 Rotation operator and Wigner D-function

Molecular eigenfunctions are normally expressed in rotated coordinates. The representations of the eigenfunction in original and rotated coordinate systems are connected with the rotation operator, $\mathcal{R}(\alpha, \beta, \gamma)$, Euler angles α, β and γ, and the Wigner D-functions,

$$\begin{aligned}\langle \mathbf{r}_1, \mathbf{r}_2, \ldots, \mathbf{r}_N\,|JM\rangle &= \sum_{\Omega=-J}^{J} \langle \mathbf{r}_1, \mathbf{r}_2, \ldots, \mathbf{r}_N\,|\mathcal{R}(\alpha, \beta, \gamma) \\ &\quad \times |J\Omega\rangle\langle J\Omega\,|\mathcal{R}^{\dagger}(\alpha, \beta, \gamma)|JM\rangle \\ &= \sum_{\Omega=-J}^{J} \langle \mathbf{r}'_1, \mathbf{r}'_2, \ldots, \mathbf{r}'_N\,|J\Omega\rangle\, D^{J*}_{M\Omega}(\alpha, \beta, \gamma).\end{aligned} \tag{E.2}$$

Angular momentum is the generator of rotations, therefore, one can expect that application of the discrete operator C_2 to the arguments of the D-function would yield a relationship between angular momentum and parity.

In terms of spatial and angular coordinates appropriate to the diatomic molecule, equation (E.2) can be written as

$$\begin{aligned}&\langle \rho, \zeta, \chi, \mathbf{r}_2, \ldots, \mathbf{r}_n, \mathrm{r}, \theta, \phi\,|nvJM\rangle \\ &= \sum_{\Omega=-J}^{J} \langle \rho, \zeta, \chi', \mathbf{r}'_2, \ldots, \mathbf{r}'_N, \mathrm{r}, \theta', \phi'\,|nvJ\Omega\rangle\, D^{J*}_{M\Omega}(\alpha, \beta, \gamma).\end{aligned} \tag{E.3}$$

Here, ρ is the distance of one electron (the electron arbitrarily labeled 1 but it could be any one of the electrons), ζ is the distance of that electron above or below the plane that passes through the center of mass of the two nuclei (the coordinate origin), and χ is the angle of rotation of that electron about the internuclear vector $\mathbf{r}(\mathrm{r}, \theta, \phi)$. The vibrational quantum number, v, has been extracted from the quantum numbers collection, n, which represents all required quantum numbers except J, M, Ω, and v.

The variables ρ, ζ, and r are scalars, unaffected by rotations. The physical rotation ϕ and the angle of coordinate rotation α are about the same axis, namely the z-axis. The physical rotation θ and the angle of coordinate rotation β are also about the same axis, namely the first intermediate y-axis of the full coordinate rotation. The angles χ and γ are both rotations about the z'-axis. Thus,

$$\phi' = \phi - \alpha, \ \theta' = \theta - \beta, \ \chi' = \chi - \gamma. \tag{E.4}$$

In coordinate rotations, one is at liberty to choose α, β and γ. If one chooses for the angles $\alpha = \phi$, $\beta = \theta$ and $\gamma = \chi$, then all angular dependence of $\langle \rho, \zeta, \chi', \mathbf{r}_2', \ldots, \mathbf{r}_N', \mathrm{r}, \theta', \phi' \,|nvJ\Omega\rangle$ is removed. This yields the Wigner and Witmer [8] diatomic eigenfunction,

$$\begin{aligned} &\langle \rho, \zeta, \chi, \mathbf{r}_2, \ldots, \mathbf{r}_n, \mathrm{r}, \theta, \phi \,|nJM\rangle \\ &= \sum_{\Omega=-J}^{J} \langle \rho, \zeta, \mathbf{r}_2', \ldots, \mathbf{r}_N', \mathrm{r} \,|nv\rangle \, D_{M\Omega}^{J^*}(\phi, \theta, \chi). \end{aligned} \tag{E.5}$$

The values of the quantum numbers, J and Ω, influence the electronic–vibrational eigenfunction $\langle \rho, \zeta, \mathbf{r}_2', \ldots, \mathbf{r}_N', r \,|nv\rangle$, but the electronic–vibrational eigenfunction is not an angular momentum state vector.

E.4 Parity of diatomic states

Parity is rotationally invariant. Inversion of the signs of all rotated coordinates inverts the signs of all un-rotated coordinates, and vice versa. Therefore, the parity operator can be represented by $C_2(x')\,\sigma_v(y', z')$. The application to the right side of the Wigner–Witmer diatomic eigenfunction (E.5) yields the parity eigenvalues,

$$p = -p_\Sigma(-1)^J \qquad J \text{ half-integer}, \tag{E.6}$$

$$p = p_\Sigma(-1)^J \qquad J \text{ integer}. \tag{E.7}$$

The constant part of the parity eigenvalue, p_Σ, is labeled in accordance with standard spectroscopic notation. The imaginary values of $(-1)^J$ occurring when J is half-integer can be avoided if one adopts the convention [3] to always subtract 1/2 from J when J is half-integer. With this convention, equation (E.6) is replaced by

$$p = -p_\Sigma(-1)^{J-1/2} \qquad J \text{ half-integer}. \tag{E.8}$$

The value of p_Σ does not depend upon quantum numbers. It is a global value applying to all states of a given molecule. If the diatomic molecule can be said to have an intrinsic parity, then it is clearly p_Σ. One would expect the product of the intrinsic parities of the fundamental particles composing the molecule to equal p_Σ.

E.5 Parity in an algorithm for computing diatomic spectra

The following describes an algorithm in which equations (E.7) and (E.8) become practical equations for computing diatomic parity.

Consider the algorithm for computation of the wavelengths and intensities in the spectrum of a molecule from the first principles of quantum mechanics. The upper H' and lower H Hamiltonian matrices are computed and numerically diagonalized by unitary matrices, U' and U. The upper $F'_{n'v'J'}$ and lower F_{nvJ} terms are the eigenvalues of the Hamiltonians,

$$F'_{n'v'J'} = U'^{\dagger} H' \; U' \tag{E.9a}$$

$$F_{nvJ} = U^{\dagger} H \; U, \tag{E.9b}$$

and the vacuum wavenumbers, $\tilde{\nu}$, of the predicted spectral lines,

$$\tilde{\nu} = F'_{n'v'J'} - F_{nvJ}, \tag{E.10}$$

are term differences. Of the very large number of computed term differences, only those for which the Condon and Shortley [9] line strength does not vanish are spectral lines. The line strength, $S(nvJ,\, n'v'J')$, is the sum over all M and M' of the irreducible tensor $T_k^{(q)}$ expectation values, $\langle nvJM \,|T_k^{(q)}| n'v'J'M' \rangle$. The exact separation of the total angular momentum in the Wigner–Witmer diatomic eigenfunction results in a diatomic line strength composed of two parts,

$$S(nvJ,\, n'v'J') = S(nv,\, n'v')\; S(J,\, J'), \tag{E.11}$$

the electronic–vibrational strength, $S(nv,\, n'v')$, and the unitless rotational line strength or Hönl–London factor, $S(J,\, J')$. The Born–Oppenheimer approximation separates the electronic–vibrational strength into electronic and vibrational parts. In Hund's case (a) basis built from the Wigner–Witmer eigenfunction, the third Euler angle, $\chi = \gamma$, appears in the Wigner D-function,

$$\begin{aligned} \langle a \rangle &= \langle nvJM\Omega S\Sigma \rangle \\ &= \sqrt{\frac{2J+1}{8\pi^2}} \langle \rho,\, \zeta,\, \mathbf{r}'_2,\, \ldots\, ,\, \mathbf{r}'_N,\, \mathrm{r} \,| nv \rangle \,| S\Sigma \rangle \, D^{J^*}_{M\Omega}(\phi,\, \theta,\, \chi). \end{aligned} \tag{E.12}$$

The algorithm for computation of the vacuum wavenumbers, $\tilde{\nu}$, and diatomic spectral line strengths, $S(nvJ,\, n'v'J')$, is a straightforward application of quantum mechanics, but except for the very simplest molecules is also very far removed from the realm of the possible. However, with two very stringent caveats, the algorithm can be implemented for the diatomic molecule. The first caveat is that the vacuum wavenumbers, $\tilde{\nu}$, for many spectral lines in many bands of a band system must have been experimentally measured with high accuracy such as that provided by Fourier transform spectroscopy. Secondly, using semi-empirical molecular constants one must be able to build upper and lower Hamiltonian matrices whose eigenvalue differences accurately predict the measured vacuum wavenumbers. A fitting process is required [2]. One assumes trial values for the molecular constants, computes the spectral lines positions, $\tilde{\nu}$, and from the differences between $\tilde{\nu} - \tilde{\nu}_{\mathrm{exp}}$ finds the corrections to the molecular parameters. The difference between computed and measured line positions will typically equal the measurement error margins.

As a specific example, the line position data of Faris and Cosby [10] are used for the NO beta (3,0) band for the purpose of creating a complete line list for the band with line strengths. Figure E1 illustrates a spectrum generated from the NO line list.

A multiphoton 1 + 1 excitation was used to observe 10 of the 12 possible branches [10], with particular attention to the parity designations of the numerous Λ doublets. These data are particularly suited for testing applications of the algorithm for the calculation of diatomic spectra. A total of 428 lines were fitted with a standard deviation of 0.030 cm^{-1}, and a line list having no missing lines for the range of upper and lower J values was computed. Table E1 provides details of the lines displayed in figure E1. The computed parity eigenvalues agree with those assigned by Faris and Cosby [10].

Hund's case (a) basis is mathematically complete. A sum of basis functions, $|a\rangle$, can be quantitatively very accurate. The parity operator, $\mathcal{P}$, commutes with the Hamiltonian. Thus, the orthogonal matrix that diagonalizes the case (a) representation of the Hamiltonian will also diagonalize the case (a) representation of $\mathcal{P}$.

The exact separation of the coordinates of the total angular momentum in the Wigner D-function greatly simplifies implementation of the algorithm which uses non-vanishing line strengths to determine if a computed term difference represents an allowed spectral line. The Hönl–London factors are computed from Hund's case (a) transition moment and the matrices, U and U', which diagonalize the upper and lower Hamiltonians.

A single selection rule handles all types of diatomic spectra. If the Hönl–London factor, $S(J, J')$, is non-vanishing, then the transition is allowed. Parity plays no part in the fitting process which determines the molecular parameters, but the parity eigenvalues are computed from the finalized values of the molecular parameters. The

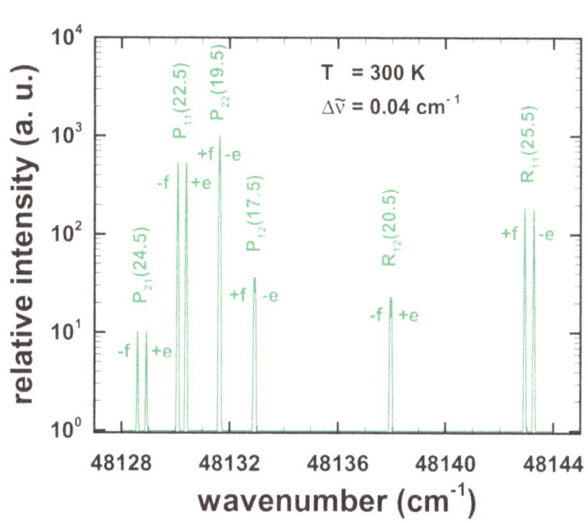

Figure E1. Section of computed spectrum of the NO $B\,^2\Pi - X\,^2\Pi$ (3,0) band. The lines are further described in table E1 [1].

Table E1. Lines of the NO $B^2\Pi - X^2\Pi$ (3,0) band, see figure E1. The Hönl–London factors, $S(J, J')$, and parity eigenvalues, p, are derived from numerical diagonalization of Hamiltonians in Hund's case (a) basis. The $P_{22}(19.5)$ Λ doublet is not resolved in the experiments [10].

J		p	$F_{J'}$ (cm^{-1})	F_J (cm^{-1})	$\tilde{\nu}$ (cm^{-1})	$S(J, J')$	$\tilde{\nu} - \tilde{\nu}_{\text{exp}}$ (cm^{-1})
24.5	P_{21}	$-f$	49 111.456	982.866	48 128.590	0.813	0.021
24.5	P_{21}	$+e$	49 111.501	982.588	48 128.914	0.812	0.078
22.5	P_{11}	$-f$	48 952.618	822.538	48 130.080	21.749	0.014
22.5	P_{11}	$+e$	48 952.673	822.282	48 130.390	21.805	−0.020
19.5	P_{22}	$+f$	48 876.118	744.504	48 131.615	18.832	−0.007
19.5	P_{22}	$-e$	48 876.146	744.505	48 131.641	18.786	
17.5	P_{12}	$+f$	48 747.194	614.306	48 132.888	0.515	−0.013
17.5	P_{12}	$-e$	48 747.244	614.307	48 132.936	0.527	−0.012
20.5	R_{12}	$-f$	48 952.618	814.699	48 137.919	0.881	0.034
20.5	R_{12}	$+e$	48 952.673	814.701	48 137.971	0.880	−0.077
25.5	R_{11}	$+f$	49 210.930	1068.021	48 142.908	25.424	0.034
25.5	R_{11}	$-e$	49 210.985	1067.731	48 143.254	25.317	−0.018

presented algorithm can be used to predict molecular spectra for the purpose of fitting measured data [11].

References

[1] Hornkohl J O and Parigger C G 2017 *Int. J. Mol. Theor. Phys.* **1** 00103

[2] Zare R N, Schmeltekopf A L, Harrop W J and Albritton D L 1973 *J. Mol. Spectrosc.* **46** 37

[3] Brown J M *et al* 1975 *J. Mol. Spectrosc.* **55** 500

[4] Hougen J T 2001 *The Calculation of Rotational Energy Levels and Rotational Line Intensities in Diatomic Molecules* National Institute of Standards and Technology http://physics.nist.gov/DiatomicCalculations Originally published as *The Calculation of Rotational Energy Levels and Rotational Line Intensities in Diatomic Molecules*, J T Hougen NBS Monograph 115 (1970)

[5] Røeggen I 1971 *Theor. Chim. Acta* **21** 398

[6] Judd B 1975 *Angular Momentum Theory for Diatomic Molecules* (New York: Academic)

[7] Larsson M 1981 *Phys. Scr.* **23** 835

[8] Wigner E and Witmer E E 1928 *Z. Phys.* **51** 859
Hettema H 2000 *Quantum Chemistry: Classic Scientific Papers* (Singapore: World Scientific), p 287

[9] Condon E U and Shortley G 1953 *The Theory of Atomic Spectra* (Cambridge: Cambridge University Press)

[10] Faris G W and Cosby P C 1992 *J. Chem. Phys.* **97** 7073

[11] Parigger C G, Woods A C, Surmick D M, Gautam G, Witte M J and Hornkohl J O 2015 *Spectrochim. Acta Part B: At. Spectrosc.* **107** 132

Appendix F

Rotational line strengths for the CN BX (5,4) band

This appendix communicates rotational line strengths for the CN $B^2\Sigma^+-X^2\Sigma^+$ (5,4) band [1]. Rotational line strengths, computed from eigenvectors of Hund's case (a) matrix representations of the upper and lower Hamiltonians using Wigner–Witmer basis functions, show a larger than expected influence from the well known perturbation in the (5,4) band. Comparisons with National Solar Observatory experimental Fourier transform spectroscopy data reveal nice agreement of measured and predicted spectra.

F.1 Introduction

The CN violet $B^2\Sigma^+-X^2\Sigma^+$ band system is one of the most studied band systems. Ram *et al* [2] and Brooke *et al* [3] have summarized the available experimental and theoretical information. Of the many known bands in the violet system, only the (5,4) band is considered here. This band exhibits a weak, quantitatively understood perturbation [4] caused by mixing of the $v = 17$ level of $A^2\Pi$ with the $v = 5$ level of $B^2\Sigma^+$. The particular perturbation of the CN (5,4) band is not considered in the study by Brooke *et al* [3] but is evaluated in this work by isolating the spectral features of this band that is part of the CN violet system. Numerical diagonalisations of upper and lower Hamiltonians with and without the perturbation are investigated and compared with available experimental spectra. The simulations rely on determining rotational strengths without parity-partitioned Hamiltonians. It is anticipated that the investigated (5,4) band modifications can be possibly confirmed with the new PGOPHER program recently released by Western [5].

F.2 CN (5,4) band spectra

For the computation of rotational spectra, the square of transition moments are numerically computed using the eigenvectors of upper and lower Hamiltonians. This

doi:10.1088/978-0-7503-1989-8ch26

approach can also be selected in the new PGOPHER program [5]. For the diatomic molecule, the results effectively yield the Hönl–London factors yet we do not utilize tabulated Hönl–London factors that are available in standard textbooks. Table F1 and figures F1 and F2 compare results obtained with and without taking into account the mixing. Results of modeling the angular momentum states of the upper $v = 5$ vibrational level as a mixture of $^2\Sigma$ and $^2\Pi$ Hund's case (a) basis functions, a so-called 'de-perturbation' or perturbation analysis, agree well with that of Ito *et al* [4] who used the line position measurements of Engleman [6]. The 100 lines of the more recent data of Ram *et al* [2] were fitted with a standard deviation of 0.025 cm^{-1}. Failure to include spin–orbit mixing of the $B^2\Sigma^+$ and $A^2\Pi$ basis states increased the standard deviation to 0.25 cm^{-1}.

The table and synthetic spectra reveal that the changes caused by spin–orbit mixing are relatively very much larger for the rotational line strengths, $S(J', J)$, than for the line positions, $\tilde{\nu}$. The simulation results compare nicely with measured spectra [2] available from the National Solar Observatory (NSO) at Kitt Peak [7]. Figure F3 displays the recorded and simulated spectra for a resolution of 0.03 cm^{-1}.

The influence of $^2\Sigma^+ + {}^2\Pi$ mixing on the rotational line strengths, $S(J', J)$, was recognized because computation of $S(J', J)$ is an integral part of the unique line

Table F1. Lines in the CN $B\ {}^2\Sigma^+ X\ {}^2\Sigma^+$ (5,4) band near the perturbation. $\tilde{\nu}$ are the fitted line positions, $S(J', J)$ are the rotational line strengths computed in the fitting algorithm. $S^{(0)}(J', J)$ and $\Delta\tilde{\nu}^{(0)}$ are the line strengths and errors in the fitted line positions, respectively, when the off-diagonal spin–orbit coupling constants $\langle AL+\rangle$ and $\langle BL+\rangle$ are set equal to 0. Spin–orbit mixing of $B^2\Sigma^+$ and $A^2\Pi$ shifts the upper e parity levels. An error in the $\tilde{\nu}(J', J)$ associated with these upper e parity levels is produced if the mixing is ignored. A relatively large fractional error (e.g. −3.974/17.455 versus −1.870/28 032 for $R_{11}(12.5)$) can occur in the rotational line strengths, $S(J', J)$.

J'	J		p'	$\tilde{\nu}$	$S_{J'J}$	$\Delta\tilde{\nu}$	$S^{(0)}_{J'J}$	$\Delta\tilde{\nu}^{(0)}$
9.5	8.5	R_{11}	$-e$	28 013.117	9.474	−0.010	9.474	0.337
9.5	8.5	R_{22}	$-f$	28 017.421	9.474	0.001	9.474	−0.059
10.5	9.5	R_{11}	$+e$	28 016.992	9.1988	−0.004	10.476	0.600
10.5	9.5	R_{22}	$-f$	28 021.651	11.171	−0.000	10.476	−0.067
11.5	10.5	R_{11}	$-e$	28 020.540	7.868	−0.041	11.478	1.193
11.5	10.5	R_{22}	$-f$	28 025.866	12.240	0.006	11.478	−0.067
12.5	11.5	R_{22}	$-f$	28 030.125	13.288	0.007	12.480	−0.072
12.5	11.5	R_{11}	$+e$	28 030.431	13.812		12.480	
13.5	12.5	R_{11}	$-e$	28 032.081	17.455	−0.053	13.481	−1.870
13.5	12.5	R_{22}	$-f$	28 034.428	14.325	0.011	13.481	−0.073
14.5	13.5	R_{11}	$+e$	28 035.672	17.919	−0.005	14.483	−1.102
14.5	13.5	R_{22}	$-f$	28 038.773	15.356	0.013	14.483	−0.076
15.5	14.5	R_{11}	$-e$	28 039.742	18.442	0.007	15.484	−0.807
15.5	14.5	R_{22}	$-f$	28 043.161	16.383	0.009	15.484	−0.084
16.5	15.5	R_{11}	$+e$	28 043.989	19.132	0.011	16.485	−0.655
16.5	15.5	R_{22}	$-f$	28 047.590	17.405	0.006	16.485	−0.091

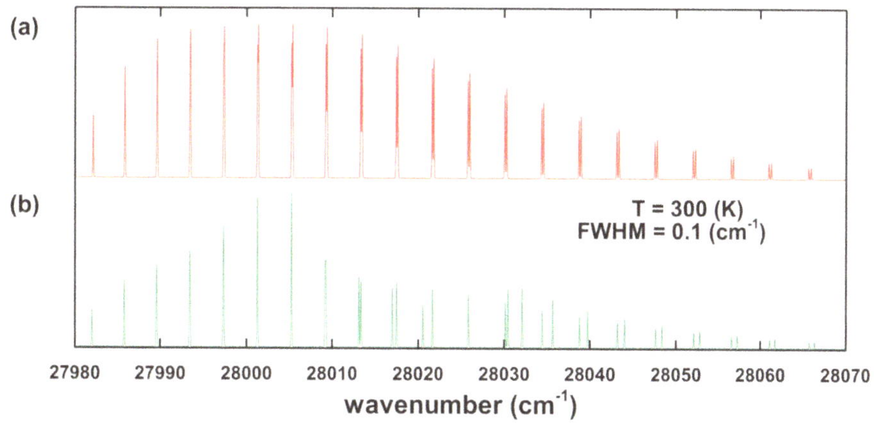

Figure F1. Synthetic emission spectra showing the influence of inclusion of the $v = 17$, $A^2\Pi$ basis in the upper $v = 5$ state of the CN violet (5,4) band. In the upper spectrum, (a), the upper states are pure $^2\Sigma^+$. The $v = 17$, $A^2\Pi$ energy eigenvalues lie very near the $v = 5$, $B^2\Sigma^+$ eigenvalues, and this explains the large influence of the $A^2\Pi$ basis. In the lower spectrum, (b), the upper states are treated as the sum $c_\Sigma\,{}^2\Sigma^+ + c_\Pi\,{}^2\Pi$ with $c_\Sigma \gg c_\Pi$. Only R branch lines are shown here, including those given in table F1 [1].

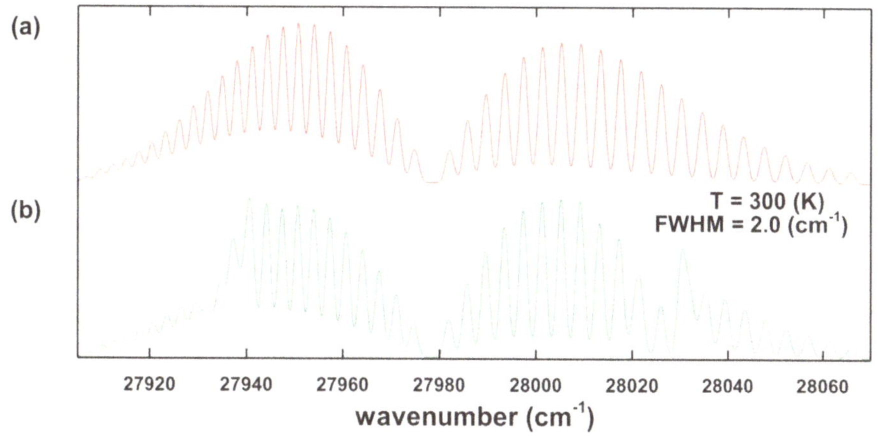

Figure F2. The lower resolution spectra include both the P and R branches. (a) pure, (b) addition of a small amount of $^2\Pi$ to the upper basis affects the lower spectrum of the violet (5,4) band even at low resolution [1].

position fitting algorithm. Upper and lower Hamiltonian matrices in Hund's case (a) basis are numerically diagonalised, and the spectral line vacuum wavenumber $\tilde{\nu}$ is the difference between upper and lower Hamiltonian eigenvalues. To determine which of the many eigenvalue differences represent allowed spectral lines, the factor $S(J', J)$ is computed from the upper and lower eigenvectors for each eigenvalue difference. A non-vanishing $S(J', J)$ denotes an allowed diatomic spectral line. Parity partitioned effective Hamiltonians are not used. Parity and branch designation are not required in the fitting algorithm. Input data to the fitting program is a table of vacuum wavenumber $\tilde{\nu}$ versus J' and J. The non-vanishing of the rotational strength is the only selection rule used. Applications of this rule leads to the

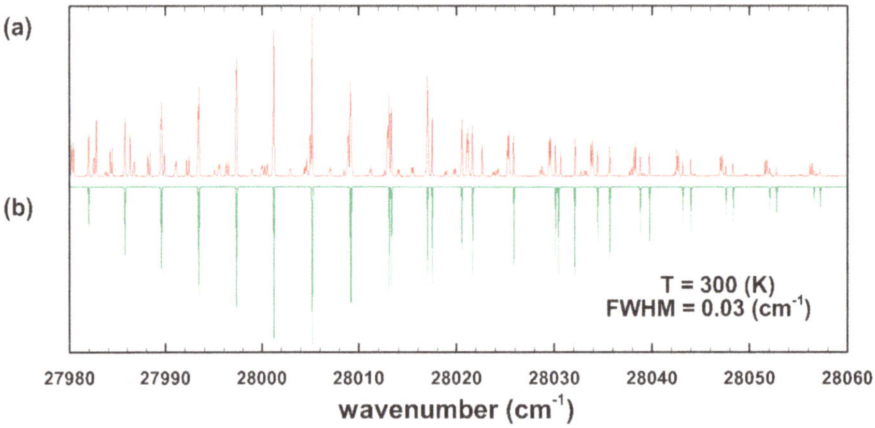

Figure F3. Comparison of measured and simulated spectra. (a) Segment of the recorded [2] Fourier transform spectrum 920212R0.005 [7], (b) computed spectrum for a temperature of 300 K and a spectral resolution of 0.03 cm^{-1}. The computed (5.4) band is flipped vertically to show how the predicted line positions of the R-branch match the vacuum wavenumbers of the experimental spectrum [1].

establishment of spectral data bases for diatomic molecular spectroscopy of selected transitions [8]. Over and above the PGOPHER program [5], there are other extensive efforts in predicting diatomic molecular spectra including for instance the so-called DUO program [9] for diatomic spectroscopy.

F.3 Wigner–Witmer diatomic eigenfunction

Hund's case (a) basis functions were derived from the Wigner and Witmer [10] diatomic eigenfunction,

$$\langle \rho, \zeta, \chi, \mathbf{r}_2, \ldots, \mathbf{r}_N, r, \theta, \phi \,|nvJM\rangle = \sum_{\Omega=-J}^{J} \langle \rho, \zeta, \mathbf{r}'_2, \ldots, \mathbf{r}'_N, r \,|nv\rangle \times D^{J^*}_{M\Omega}(\phi, \theta, \chi). \tag{F.1}$$

The coordinates are ρ the distance of one electron (the electron arbitrarily labeled 1 but it could be any one of the electrons) from the internuclear vector $\mathbf{r}(r, \theta, \phi)$, the distance ζ of that electron above or below the plane perpendicular to $\mathbf{r}$ and passing through the center of mass of the two nuclei (the coordinate origin), the angle χ for rotation of that electron about the internuclear vector $\mathbf{r}$, and the remaining electronic coordinates $\mathbf{r}_2, \ldots, \mathbf{r}_N$ in the fixed and $\mathbf{r}'_2, \ldots, \mathbf{r}'_N$ in the rotating coordinate system. The vibrational quantum number v has been extracted from the quantum numbers collection n which represents all required quantum numbers except J, M, Ω, and v. The Wigner–Witmer diatomic eigenfunction has no application in polyatomic theory, but for the diatomic molecule the exact separation of the Euler angles is a clear advantage over the Born–Oppenheimer approximation for the diatomic molecule in which the angle of electronic rotation, χ, is unnecessarily separated from the angles describing nuclear rotation, θ and ϕ. Equation (F.1) can be derived by writing the general equation for coordinate (passive) rotations α, β, and γ of the eigenfunction, replacing two generic coordinate vectors with the diatomic vectors $\mathbf{r}(r, \theta, \phi)$ and $\mathbf{r}'(\rho, \zeta, \chi)$, and equating the angles of coordinate rotation to the angles of physical rotation ϕ, θ, and ϕ. The general equation for coordinate rotation holds in isotropic space, and therefore the quantum numbers J, M, and Ω in the Wigner–Witmer eigenfunction include all electronic and nuclear spins. If nuclear spin were to be included, J, M, and Ω would be replaced by F, M_F, and Ω_F, but hyperfine structure is not resolved in the (5,4) band data reported by [2], and equation (F.1) is written with the appropriate spectroscopic quantum numbers.

It is worth noting that the rotation matrix element $D^{J}_{M\Omega}(\phi, \theta, \chi)$ and its complex conjugate $D^{J*}_{M\Omega}(\phi, \theta, \chi)$ do not fully possess the mathematical properties of quantum mechanical angular momentum. It is well known that a sum of Wigner D-functions is required to build an angular momentum state. The equation

$$J'_{\pm}\, D^{J*}_{M\Omega}(\phi, \theta, \chi) = \sqrt{J(J+1) - \Omega(\Omega \mp 1)}\; D^{J*}_{M,\Omega\mp 1}(\phi, \theta, \chi) \tag{F.2}$$

is not a phase convention [11–13] but a mathematical result readily obtained from equation (F.1) and

$$J'_{\pm}\, |J\Omega\rangle = \sqrt{J(J+1) - \Omega(\Omega \pm 1)}\; |J, \Omega \pm 1\rangle, \tag{F.3}$$

in which the prime on the operator $J'_{\pm}$ indicates that it is written in the rotated coordinate system where the appropriate magnetic quantum number is Ω.

F.4 Hund's basis functions

Hund's case (a) basis function based upon the Wigner–Witmer diatomic eigenfunction is

$$\begin{aligned} |a\rangle &= \langle \rho, \zeta, \chi, \mathbf{r}'_2, \ldots, \mathbf{r}'_N, r, \theta, \phi\, | nvJMS\Lambda\Sigma\Omega\rangle \\ &= \sqrt{\frac{2J+1}{8\pi^2}}\, \langle \rho, \zeta, \mathbf{r}'_2, \ldots, \mathbf{r}'_N, r\, | nv\rangle\, |S\Sigma\rangle\, D^{J*}_{M\Omega}(\phi, \theta, \chi). \end{aligned} \tag{F.4}$$

As noted above, a sum of $|a\rangle$ basis functions is required to build an eigenstate of angular momentum. The basis function would also not be an eigenstate of the parity operator. The case (a) matrix elements, $p^{(a)}_{ij}$, of the parity operator $\mathcal{P}$,

$$p^{(a)}_{ij} = p_{\Sigma}(-)^{J}\delta(J_iJ_j)\,\delta(\Omega_i, -\Omega_j)\,\delta(\Lambda_i, -\Lambda_i)\,\delta(n_in_j), \tag{F.5}$$

show that a single $|a\rangle$ basis function is not an eigenstate of parity. The procedure called parity symmetrization adds $|JM\Omega\rangle$ and $|JM, -\Omega\rangle$ basis functions thereby destroying the second magnetic quantum number Ω and yielding a function which at least possesses the minimal mathematical properties of an eigenstate of angular momentum, parity, and the other members of the complete set of commuting operators. The general procedure would be to continue adding basis functions to the upper and lower bases until eigenvalue differences between the upper and lower Hamiltonians accurately predict measured line positions.

Table F2. Molecular parameters used in this work that relies on Hamiltonians that are not partity-partitioned. Values not followed by a number in parenthesis were held fixed or an error estimate was not computed. A value in parenthesis is the standard deviation in the fitted value. Parameters for the $A\,^2\Pi$ state were fitted by the Nelder–Mead minimization algorithm using values given by Brooke *et al* [3] as trial values. Error estimates were not computed, and the values of Brooke *et al* [3] were only very slightly changed.

	$X^2\Sigma^+$ $v = 4$	$B^2\Sigma^+$ $v = 5$	$A^2\Pi$ $v = 17$
B_v	1.820 866(13)	1.845 727(13)	1.404 833
D_v	$6.172(36) \times 10^{-6}$	$8.003(38) \times 10^{-6}$	5.66×10^{-6}
A_v			−50.5253
γ_v	$-1.98(43) \times 10^{-4}$	$-1.921(44) \times 10^{-2}$	
γ_{D_v}	$-1.98(43) \times 10^{-4}$		
T_v	8011.7871	35 990.1780(25)	36 010.5732
$\langle AL+\rangle$		4.25(0.03)	
$\langle BL+\rangle$		0.0205(0.001)	

F.5 The upper Hamiltonian matrix for the (5,4) band

Electronic spin **S** interactions with electronic orbital momentum **L** and nuclear orbital momentum **R** produce both diagonal and off-diagonal matrix elements in Hund's case (a) representation of the Hamiltonian. The off-diagonal elements connect different basis states. For example, both of the mentioned spin–orbit interactions connect $^2\Sigma^+$ and $^2\Pi$. Because Van Vleck transformed Hamiltonians are not used, the appropriate parameters for the strength of these interactions are $\langle AL+\rangle$ and $\langle BL+\rangle$. Table F2 lists the molecular parameters used in the Hamiltonian matrices. Tables F3 and F4 show the Hamiltonian matrices without and with spin–orbit interactions, respectively.

F.6 A diatomic line position fitting algorithm

A basic tool for the diatomic spectroscopist is a computer program that accepts a table of experimentally measured vacuum wave numbers $\tilde{\nu}_{\rm exp}$ versus J' and J, and outputs a set of molecular parameters with which one can reproduce the $\tilde{\nu}_{\rm exp}$ with a standard deviation comparable to the estimated experimental error. In practice, an experimental line list frequently shows gaps, viz. spectral lines are missing. Following a successful fitting process, one can use the molecular parameters to predict all lines. A computed line list is especially useful when it includes the Condon and Shortley [15] line strength from which the Einstein coefficients and oscillator strength [16, 17] and the HITRAN line strength [18] can be calculated. A feature of the line fitting program described below is its use of non-zerorotational strengths (see equation (F.8) below) to mark which of the many computed differences between upper and lower term values represents the vacuum wavenumber of an allowed spectral line. Consequently, the fitting process creates a complete line list including rotational factors. Parity plays no part in the fitting process, but the same orthogonal

Table F3. Hamiltonian matrix for states modeled as the mixture of $^2\Sigma^+$ and $^2\Pi$ basis states. Off-diagonal spin–orbit coupling has been removed. Consequently, the 2×2 matrices along the main diagonal are independent, and could be individually diagonalised. The bottom row contains the energy eigenvalues. Using matrices like these to model upper states of the CN violet (5,4) band, the 100 experimental spectral lines reported by Ram *et al* [2] were fitted with a standard deviation of 0.25 cm^{-1}. This Hamiltonian was computed for $\langle AL+\rangle = \langle BL+\rangle = 0$ but is otherwise identical to the Hamiltonian in table F4. Standard Hund's case (a) matrix elements [11, 13] were used.

			v	5	5	17	17	17	17
			Λ	0	0	−1	−1	1	1
			Σ	−0.5	0.5	−0.5	0.5	−0.5	0.5
v	Λ	Σ	Ω	−0.5	0.5	−1.5	−0.5	0.5	1.5
5	0	−0.5	−0.5	36 351.6409	−25.6707	0	0	0	0
5	0	0.5	0.5	−25.6707	36 351.6409	0	0	0	0
17	−1	−0.5	−1.5	0	0	36 257.6340	−19.5866	0	0
17	−1	0.5	−0.5	0	0	−19.5866	36 310.9646	0	0
17	1	−0.5	0.5	0	0	0	0	36 310.9646	−19.5866
17	1	0.5	1.5	0	0	0	0	−19.5866	36 257.6340
		E_{nvJ}	36 377.3116	36 325.9702	362 51.2135	36 317.3851	36 317.3851	36 251.2135	

Table F4. Off-diagonal spin–orbit coupling 6 × 6 matrix. From the three independent 2 × 2 matrices of table F3, the off-diagonal matrix elements mix Hund's case (a) basis states, and the standard deviation of the spectral line fit mentioned in table F3 is reduced by a factor of 10 to 0.025 cm^{-1}. The spin–orbit coupling constants $\langle AL+\rangle = 4.25(0.03)$ and $\langle BL+\rangle = 0.205(0.001)$ were used in computation of this Hamiltonian. This single 6 × 6 matrix describing $^2\Pi - ^2\Sigma^+$ mixing can be compared with the two 3 × 3 parity partitioned matrices of Brown and Carrington [14].

			v	5	5	17	17	17	17
			Λ	0	0	−1	−1	1	1
			Σ	−0.5	0.5	−0.5	0.5	−0.5	0.5
v	Λ	Σ	Ω	−0.5	0.5	−1.5	−0.5	0.5	1.5
5	0	−0.5	−0.5	36 351.6409	−25.6707	2.8566	2.3274	2.8639	0
5	0	0.5	0.5	−25.6707	36 351.6409	0	2.8639	2.3274	2.8566
17	−1	−0.5	−1.5	2.8566	0	36 257.6340	−19.5866	0	0
17	−1	0.5	−0.5	2.3274	2.8639	−19.5866	36 310.9646	0	0
17	1	−0.5	0.5	2.8639	2.3274	0	0	36 310.9646	−19.5866
17	1	0.5	1.5	0	2.8566	0	0	−19.5866	36 257.6340
		E_{nvJ}	36 377.3957	36 327.7869	36 250.9625	36 317.3525	36 315.8194	36 251.1620	

matrix that diagonalises the case (a) Hamiltonian matrix will also diagonalise the case (a) parity matrix whose elements are given in equation (F.5). The $p = \pm 1$ parity eigenvalue becomes a computed quantity, and the *e*/*f* parity designation is established from the parity eigenvalue using the accepted convention of Brown *et al* [19].

Trial values of upper and lower state molecular parameters, typically taken from previous works by others for the band system in question, are used to compute upper H' and lower H Hamiltonian matrices in the case (a) basis given by equation (F.4) for specific values of J' and J. The upper and lower Hamiltonians are numericallydiagonalised,

$$T' = \tilde{U}' H' U' \tag{F.6a}$$

$$T = \tilde{U} H U \tag{F.6b}$$

giving the upper T' and lower T term values. The vacuum wavenumber $\tilde{\nu}$ is determined,

$$\tilde{\nu}_{ij} = T'_i - T_j, \tag{F.7}$$

and the rotational strength is evaluated,

$$S_{ij}(J', J) = (2J + 1) \left| \sum_n \sum_m \tilde{U}'_{in} \langle J\Omega; q, \Omega' - \Omega | J'\Omega' \rangle U_{mj} \, \delta(\Sigma'_n \Sigma_m) \right|^2 . \tag{F.8}$$

The degree of the tensor operator, q, responsible for the transitionsamounts to $q = 1$ for electric dipole transitions. For a non-zero rotational factors, $S(J', J)$, the vacuum wavenumber $\tilde{\nu}_{ij}$ is added to a table of computed line positions to be compared with the experimental list $\tilde{\nu}_{\text{exp}}$ versus J' and J. The Clebsch–Gordan coefficient, $\langle J\Omega; q, \Omega' - \Omega | J'\Omega' \rangle$, is the same one appearing in the pure case (a)—case (a) formulae for $S(J', J)$. For specific values of J' and J, one constructs tables for $\tilde{\nu}_{\text{exp}}$ and computed $\tilde{\nu}_{ij}$. The errors $\Delta\tilde{\nu}_{ij}$,

$$\Delta\tilde{\nu}_{ij} = \tilde{\nu}_{ij} - \tilde{\nu}_{\text{exp}}, \tag{F.9}$$

are computed where each $\tilde{\nu}_{ij}$ is the one that most closely equals one of the $\tilde{\nu}_{\text{exp}}$. Once values of $\tilde{\nu}_{ij}$ and $\tilde{\nu}_{\text{exp}}$ are matched, each is marked unavailable until a new list of $\tilde{\nu}_{ij}$ is computed. The indicated computations are performed for all values of J' and J in the experimental line list, and corrections to the trial values of the molecular parameters are subsequently determined from the resulting $\Delta\tilde{\nu}_{ij}$. The entire process is iterated until the parameter corrections become negligibly small. As this fitting process successfully concludes, one obtains a set of molecular parameters that predict the measured line positions $\tilde{\nu}_{\text{exp}}$ with standard deviations that essentially equal the experimental estimates for the accuracy of the $\tilde{\nu}_{\text{exp}}$.

F.7 Discussion

The influence on intensities in the (5,4) band of the CN violet system caused by the weak spin–orbit mixing, figures F1 and F2, is significantly larger than initially anticipated. This was noticed because computation of the rotational strengths is an integral part of our line position fitting program. The eigenvectors that diagonalise the Hamiltonian to yield fitted line position $\tilde{\nu}$ also yield $S(J', J)$. The eigenvectors of Van Vleck transformed effective Hamiltonians are rarely discussed in the literature, equally, the calculations of Hönl–London factors appear problematic. In established diatomic molecular practice, Hönl–London factors are determined independently of line positions. Analytical approximations utilize the parameter $Y = A/B$ to account for the influence of spin–orbit interaction on $S(J', J)$. Kovács [20] gives many examples, Li *et al* [21] give a more recent application. These analytical approximations can accurately account for intermediate spin–orbit coupling which smooth transitions between case (a) and case (b) with increasing J' and J, but show limited sensitivity to abrupt changes in $S(J', J)$ near perturbations such as those seen the (5,4) band in the CN violet system.

It is noted in passing that the $S^{(0)}(J', J)$ in table F1 that describe rotational strengths without off-diagonal spin–orbit constants, are exactly equal to pure case (b) Hönl–London factors for $^2\Sigma^+ - {}^2\Sigma^+$ transitions even though all computations were carried out in Hund's case (a) basis. This observation merely makes the point that the Hund's case (b) is defined due to the physical absence of spin–orbit coupling.

F.8 Conclusion

The Wigner–Witmer diatomic eigenfunction makes it possible to form an exact, mathematical connection between computation of $\tilde{\nu}$ and $S(J', J)$ in a single algorithm. The concept of the non-vanishing rotational strengths as the omnipotent selection rule initially conceived as a simplifying convenience in a computer algorithm is now seen to be more valuable, as evidenced in this work's analysis of the CN (5,4) band perturbations by isolating a specific branch. Future work is planned for comparisons of the CN (10,10) band spectra that include perturbation and that show promising agreements with experiments and PGOPHER predictions.

References

[1] Hornkohl J O and Parigger C G 2017 *Int. J. Mol. Theor. Phys.* **1** 00102

[2] Ram R S, Davis S P, Wallace L, Englman R, Appadoo D R T and Bernath P F 2006 *J. Mol. Spectrosc.* **237** 225

[3] Brooke J S A, Ram R S, Western C M, Li G, Schwenke D W and Bernath P F 2014 *J. Quant. Spectrosc. Radiat. Transfer*, Astrophys. J. Supp. Series **210** 23

[4] Ito H, Fukuda Y, Ozaki Y, Kondow T and Kuchitsu K 1987 *J. Mol. Spectrosc.* **121** 84

[5] Western C M 2017 *J. Quant. Spectrosc. Radiat. Transfer* **186** 221

[6] Engleman R 1974 *J. Mol. Spectrosc.* **49** 106

[7] Kroto H W 2016 National Solar Observatory (NSO) at Kitt Peak, McMath-Pierce Fourier Transform Spectrometer (FTS) Data (accessed December 20, 2016) ftp://vso.nso.edu/FTS_cdrom/FTS30/920212R0.005

[8] Parigger C G, Woods A C, Surmick D M, Gautam G, Witte M J and Hornkohl J O 2015 *Spectrochim. Acta Part B: At. Spectrosc.* **107** 132

[9] Yurchenko S N, Lodi L, Tennyson J and Stolyarov A V 2016 *Comput. Phys. Commun.* **202** 262

[10] Wigner E and Witmer E E 1928 *Z. Phys.* **51** 859
Hettema H 2000 *Quantum Chemistry: Classic Scientific Papers* (Singapore: World Scientific), p 287

[11] Zare R N, Schmeltekopf A L, Harrop W J and Albritton D L 1973 *J. Mol. Spectrosc.* **46** 37

[12] Howard B J and Brown J M 1976 *J. Mol. Phys.* **31** 1517

[13] Lefebvre-Brion H and Field R W 2004 *The Spectra and Dynamics of Diatomic Molecules* (New York: Elsevier/Academic)

[14] Brown J M and Carrington A 2003 *Rotational Spectroscopy of Diatomic Molecules* (Cambridge: Cambridge University Press)

[15] Condon E U and Shortley G 1953 *The Theory of Atomic Spectra* (Cambridge: Cambridge University Press)

[16] Hilborn R C 1982 *Am. J. Phys.* **50** 982 http://arxiv.org/ftp/physics/papers/0202/0202029.pdf

[17] Thorne A 1988 *Spectrophysics* 2nd edn (London: Chapman and Hall)

[18] Rothman L S *et al* 1998 *J. Quant. Spectrosc. Radiat. Transfer* **60** 665

[19] Brown J M *et al* 1975 *J. Mol. Spectrosc.* **55** 500

[20] Kovács I 1969 *Rotational Structure in the Spectra of Diatomic Molecules* (New York: Elsevier)

[21] Li G, Harrison J J, Ram R S, Western C M and Bernath P F 2012 *J. Quant. Spectrosc. Radiat. Transfer* **113** 67

IOP Publishing

Quantum Mechanics of the Diatomic Molecule with Applications

Christian G Parigger and James O Hornkohl

Appendix G

Intrinsic parity of the diatomic molecule

This appendix discusses appearances of selected homonuclear spectra. In quantum field theory the parity of a state is broken into two parts, one depending on the angular momentum, and the second part, called the *intrinsic parity*, which is independent of angular momentum. Particles currently viewed as fundamental (e.g. the photon and electron) are assigned an intrinsic parity, and because parity is multiplicative the intrinsic parity of a collection of fundamental particles is the product of the individual intrinsic parities. Thus, nucleons have an intrinsic parity of +1, and the intrinsic parity of an atomic or molecular system having N electrons simply $(-)^N$. Does intrinsic parity have any relevance in atomic and molecular physics? By comparing experimental spectra recorded by others with synthetic spectra computed by us using intrinsic parity $(-)^N$ in the computations, we conclude that intrinsic parity is relevant to the spectra of homonuclear diatomic molecules.

Unresolved hyperfine structure (unresolved nuclear spin states) of a diatomic molecule state effectively give the state a degeneracy factor (statistical weight) $g = (2I_a + 1)(2I_b + 1)$ where I_a is the nuclear spin of one nucleus and I_b the nuclear spin of the second nucleus. When the two nuclei are identical, the total nuclear spin statistical weight remains $(2I + 1)^2$ where $I = I_a = I_b$, but the number of nuclear spin states g_+ for which the parity is +1 and the number of states g_- for which the parity is −1 are unequal. This leads to the alternation of intensity in the spectrum of a homonuclear diatomic molcule. The experimental spectra given in the bottom trace in figures (G1)–(G3) are examples. In each figure, first above the experimental spectrum is a computed diatomic spectrum which agrees agrees reasonably well with the experimental spectrum, and the top trace in each figure is another computed spectrum which is in qualitative disagreement with the experimental spectrum. The only difference between the two computed spectra are the parity values used in their computation.

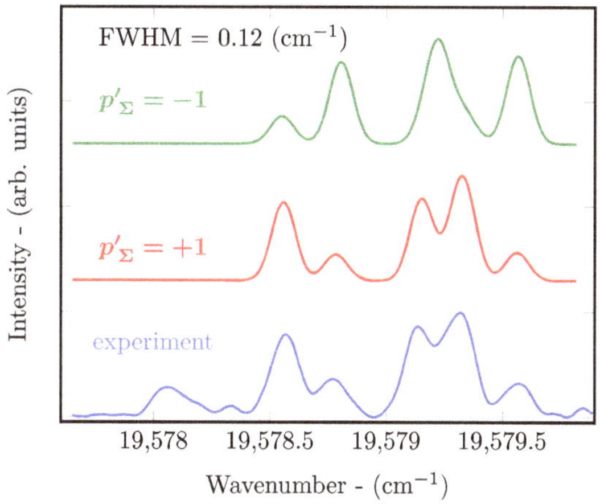

Figure G1. Two synthetic $^{13}C_2$ $d^3\Pi_g$ – $a^3\Pi_u$ spectra compared with an experimental spectrum [1]. The Λ-doublets are partially resolved. The intrinsic parity of an neutral diatomic molecule is +1 (i.e. the number of electrons is even). For the upper $d^3\Pi_g$ state, $p'_{gu} = +1$ while for the lower $a^3\Pi_u$ state, $p_{gu} = -1$. Thus, $p'_\Sigma = p_0\, p'_{gu} = +1$ and $p_\Sigma = p_0\, p_{gu} = -1$. The molecular parameters were found by fitting the line position data of [1].

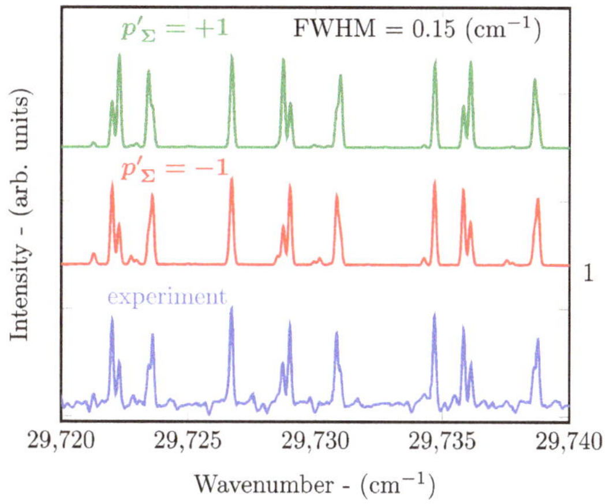

Figure G2. Two synthetic $C^3\Pi_u$ – $B^3\Pi_g$ N_2 second positive system spectra compared with an experimental spectrum [2]. The Λ-doublets are partially resolved. This spectrum reverses the situation of figure G1. The intrinsic parity is again +1, but here $p'_{gu} = -1$ and $p_{gu} = +1$, and one finds $p'_\Sigma = p_0\, p'_{gu} = -1$ and $p_\Sigma = p_0\, p_{gu} = +1$. The molecular parameters were found by fitting the line position data [2].

The synthetic spectra in figures (G1)–(G3) were computed using the following algorithm.

- For given upper total angular momentum quantum J' and lower quantum number J, the upper and lower Hamiltonian matrices H' and H are computed in Hund's case a basis and numerically diagonalized to obtain the eigenvalues F' and F and eigenvectors U' and U.

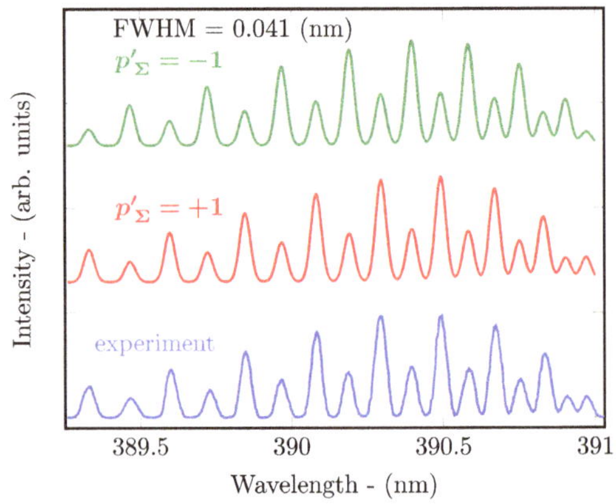

Figure G3. There are no Λ-doublets in a $^2\Sigma_u^+ - {}^2\Sigma_g^+$ spectrum, but the parity alternates from line to line causing an alternation of intensity. The two synthetic spectra are compared with an experimental spectrum [3]. The intrinsic parity of the singly ionized N_2^+ molecule is −1. Thus, $p'_\Sigma = p_0\, p'_{gu} = 1$ and $p_\Sigma = p_0\, p_{gu} = +1$. The molecular parameters were found by fitting the line position data of [4].

$$F' = U'^{\dagger} H' U' \tag{G.1}$$

$$F = U^{\dagger} H U \tag{G.2}$$

- The Hönl–London factors $S_{J'J}$ are computed,

$$S_{J'J} = (2J+1)\left| \sum_{a'} \sum_{a} \langle J\, | a \rangle \left\langle a \middle| C^{J'\Omega'}_{J\Omega q\kappa} \middle| a' \right\rangle \langle a'\, | J' \rangle \right|^2 \tag{G.3}$$

  and the term difference (i.e. line position) is computed for each non-vanishing Hönl–London factor,

$$\tilde{\nu} = F' - F \quad \text{if } S_{J'J} \neq 0 \tag{G.4}$$

- A line list is created by repeating the above for all required J' and J.
- A thermal distribution of excited states and a Gaussian spectrometer slit function having a specified full-width at half-maximum (FWHM) are assumed.
- The spectrum was broken into a number of pixels, and the contribution of each line in the computed line list to each pixel is computed.

The matrix elements of parity in Hund's case (a) basis are the product of a constant p_Σ and $(-)^J$,

$$p_{ij} = p_\Sigma\, (-)^J\, \delta_{J_i J_j}\, \delta_{\Omega_i, -\Omega_j}\, \delta_{\Lambda_i, -\Lambda_j}\, \delta_{n_i n_j}. \tag{G.5}$$

The quantum number n denotes the electronic basis, and other symbols have the standard meanings given them in diatomic spectroscopy. Parity is a member of the complete set of commuting observables (CSCO). Therefore, the orthogonal matrices U' and U that diagonalize the upper and lower Hamiltonians computed in the case (a) basis will also diagonalize the upper and lower parity matrices computed in the case (a) basis. The only difference between the two computed spectra in each of figures (G1)–(G3) are the values of the upper p'_Σ and lower p_Σ. When computing the synthetic spectra given in figures (G1)–(G3), we imposed the selection rule $p_\Sigma \neq p'_\Sigma$. Trial and error quickly finds one of two possibilities, but having a theoretical reason for choosing the value of p_Σ for a diatomic state is clearly desirable. Equation (G.5) shows a breaking of parity into two parts, p_Σ and $(-)^J$. In quantum field theory, parity is similarly broken into two parts, one depending upon parity and the other independent of parity and called the *intrinsic parity*. Intrinsic parity is believed to be a characteristic property of a fundamental particle. For example, the electron has been assigned the intrinsic parity −1. Parity is multiplicative, and the intrinsic parity of a composite particle is the product of the intrinsic parities of particles composing it. For example, nucleons have an intrinsic parity of +1. An atom or molecule has an intrinsic parity of $(-)^N$ where N is the number of electrons. Ionization and electron attachment can obviously change the intrinsic parity of a molecule, but standard electromagnetic transitions cannot. The selection rule $p'_\Sigma \neq p_\Sigma$ used in the computation of the spectra shown in figures (G1)–(G3) eliminates consideration of p_Σ as intrinsic parity. However, the value of p_Σ appears to be correctly given as the product of the intrinsic parity eigenvalue p_0 and the *gerade/ungerade* parity eigenvalue p_{gu},

$$p_\Sigma = (-)^N p_{gu} \tag{G.6}$$

$$= p_0\, p_{gu}. \tag{G.7}$$

Based upon the very limited number of comparisons with experiment given in the attached figures, we tentatively conclude that equation (G.7) describes how intrinsic parity influences the spectrum of a homonuclear diatomic molecule.

As long as the weak force is excluded from consideration, none will doubt that the unitary transformation that diagonalizes the Hamiltonian matrix will also diagonalize the parity matrix, but equation (G.5) is new and its derivation is required. The derivation of Hund's case a matrix representation of parity is based upon the Wigner–Witmer diatomic eigenfunction, not the Born–Oppenheimer approximation, and thus represents a radical departure from current diatomic theory. The considered Hamiltonian matrices are not effective Hamiltonians but are simply composed of case (a) matrix elements. Furthermore, the Hamiltonians are not separated into positive and negative parity sub-matrices, and Van Vleck reductions of the matrix dimensions are not performed.

References

[1] Amiot C 1983 *Astrophys. J. Supp. Series* **52** 329
[2] Roux F, Michaud F and Vervloet M 1989 *Can. J. Phys.* **67** 143
[3] Lazdinis S S and Carpenter R F 1973 *J. Chem. Phys.* **59** 5203
[4] Michaud F, Roux F, Davis S P, Nguyen A D and Laux C O 2000 *J. Mol. Spectrosc.* **203** 1

Appendix H

Review of diatomic laser-induced breakdown spectroscopy

This appendix summarizes diatomic molecular spectroscopy applications in laser-induced breakdown spectroscopy (LIBS) that were recently communicated in a line shape conference [1]. Moreover, this appendix also serves as a summary on how to apply the diatomic spectroscopy computations in part I for analysis of measured emission spectra.

In principle, an atomic or molecular spectrum would be computed as follows: upper and lower Hamiltonians would be enumerated in a complete basis, and numerically diagonalized to give the upper and lower energy eigenvalues and eigenvectors. The transition moments for the appropriate operator, e.g. the electric dipole transition moments, would be evaluated from the eigenvectors. The vacuum wavenumbers $\tilde{\nu}$, i.e. energy eigenvalue differences, would be found for all non-vanishing transition moments. And the line strengths for each spectral line of wavenumber $\tilde{\nu}$ would be determined as a sum of the squares of the transition moments over all transitions producing the same $\tilde{\nu}$. A line list that includes line strengths would be generated by repeating the above computations over the required range of upper and lower total angular momentum quantum numbers. The spectrum from $\tilde{\nu}_{\rm min}$ to $\tilde{\nu}_{\rm max}$ would be separated into a number of pixels, subsequently, the contribution of each line to each pixel is calculated using the line list. This appendix reviews how this algorithm can be implemented for a diatomic spectrum if the required molecular parameters are available.

H.1 Introduction

This work reports our algorithm used to compute the synthetic spectra of diatomic molecules. Figure H1 shows a measured spectrum of interest in LIBS [1–14], and it also shows two computed C_2 Swan spectra fitted to the experimental spectrum. With one caveat, the algorithm is the textbook method for computation of a spectrum.

The caveat is that a comprehensive set of accurate diatomic parameters is required for its implementation. We describe our algorithm and give an explanation of why it is based on the Wigner–Witmer diatomic eigenfunction [15] and not the Born–Oppenheimer approximation [16].

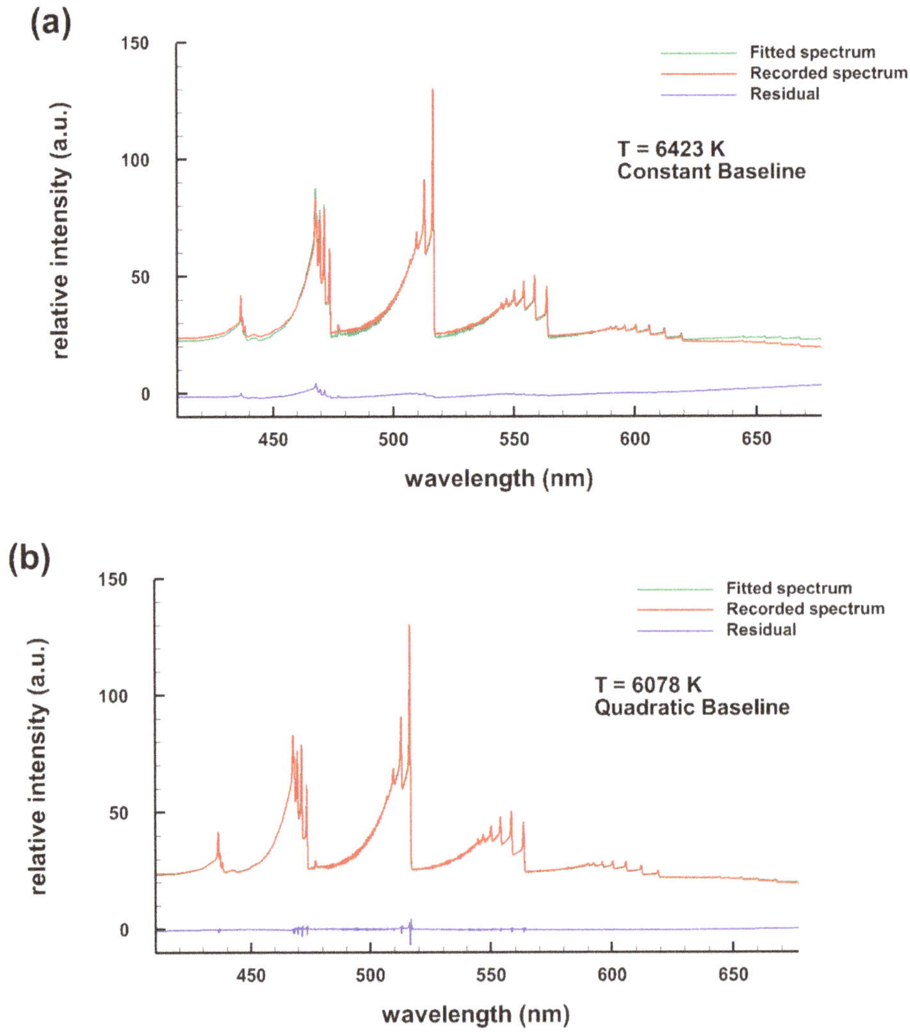

Figure H1. A recorded C_2 Swan spectrum [17] compared with two computed Swan spectra containing (a) a small constant and (b) a quadratic baseline. The temperature and baseline values were obtained by trial and error implemented with the Nelder–Mead extremum algorithm by varying the temperature for here a thermal distribution of excited states [1].

H.2 Computation of a diatomic spectrum

Except for the baseline corrections, the synthetic spontaneous emission spectra in figure (H1) were computed using the following algorithm:

- For given upper total angular momentum quantum number J' and lower quantum number J, the upper and lower Hamiltonian matrices H' and H are computed in Hund's case (a) basis and numerically diagonalized to obtain the eigenvalues F' and F and eigenvectors U' and U,

$$F' = U'^{\dagger} H' U', \tag{H.1}$$

$$F = U^{\dagger} H U. \tag{H.2}$$

- The Hönl–London line-strength factors $S_{J'J}$ are computed [18],

$$S_{J'J} = (2J+1)\left| \sum_{a'} \sum_{a} \langle J \, | a \rangle \left\langle a \middle| C^{J'\Omega'}_{J\Omega q\kappa} \middle| a' \right\rangle \langle a' \, | J' \rangle \right|^2 \tag{H.3}$$

  and the term difference, i.e. line position, is computed for each non-vanishing Hönl–London line-strength factor,

$$\tilde{\nu} = F' - F \qquad \text{if } S_{J'J} \neq 0. \tag{H.4}$$

- A line list which includes line strengths is created by repeating the above for all required J' and J.
- A thermal distribution of excited states and a Gaussian spectrometer slit function having a specified full-width at half-maximum (FWHM) are assumed.
- The spectrum is separated into a number of pixels, and the contribution of each line in the computed line list to each pixel is computed.

Use of the above algorithm requires the Wigner–Witmer diatomic eigenfunction,

$$\langle \mathbf{r}_1, \mathbf{r}_2, \ldots, \mathbf{r}_N, \mathbf{r} \, | nvJM \rangle = \sum_{\Omega=-J}^{J} \langle \rho, \zeta, \mathbf{r}'_2, \ldots, \mathbf{r}'_N, r \, | nv \rangle \, D^{J*}_{M\Omega}(\phi, \theta, \chi), \tag{H.5}$$

in which the total angular momentum states are *exactly* separated instead of a Born–Oppenheimer eigenfunction in which segregation of electronic and nuclear coordinates is enforced, thereby preventing separation of the total angular momentum. The quantum numbers in equation (H.5) are the vibrational quantum number v, the total angular momentum quantum number J, the magnetic quantum number M for the z-component of $\mathbf{J}$, the magnetic quantum number Ω for the z' component of $\mathbf{J}$, and n which is a symbol representing all other required quantum numbers and labels for continuum indices. In equation (H.5), primed and unprimed coordinates are related by coordinate rotation,

$$\begin{bmatrix} x' \\ y' \\ z' \end{bmatrix} = \mathcal{D}(\phi, \theta, \chi) \begin{bmatrix} x \\ y \\ z \end{bmatrix}. \tag{H.6}$$

The Euler angles ϕ, θ, and χ and the coordinate rotation matrix $\mathcal{D}(\phi, \theta, \chi)$ used here,

$$\mathcal{D}(\phi, \theta, \chi) = \begin{bmatrix} \cos\phi\cos\theta\cos\chi - \sin\phi\sin\chi & \sin\phi\cos\theta\cos\chi + \cos\phi\sin\chi & -\sin\theta\cos\chi \\ -\cos\phi\cos\theta\sin\chi - \sin\phi\cos\chi & -\sin\phi\cos\theta\sin\chi + \cos\phi\cos\chi & \sin\theta\sin\chi \\ \cos\phi\sin\theta & \sin\phi\sin\theta & \cos\theta \end{bmatrix} \tag{H.7}$$

are those used almost universally in the quantum theory of angular momentum [19–25]. Kovács [26] uses a different set of Euler angles. Primes do not appear on ρ (the distance of one of the electrons from the internuclear axis), ζ (this distance of this electron above or below the plane perpendicular to the internuclear axis and passing through the origin at the center of mass of the nuclei), and r (the internuclear distance) because they are scalars whose value is the same in all coordinate systems. The polar coordinates r, θ, and ϕ are those of the fictitious particle of reduced mass of the two nuclei whose motion has replaced their motion. There are N electrons and their coordinates are labeled 1, 2, ..., N. The cylindrical coordinates ρ, χ, and ζ are those of one of the electrons. In accordance with standard practice in angular momentum theory, the quantum numbers J, M, and Ω refer to the *total* angular momentum. The spectroscopic quantum numbers F, M_F, and Ω_F replace J, M, and Ω in equation (H.5) when it is written in spectroscopic notation.

In the Wigner–Witmer diatomic eigenfunction, the Euler angles ϕ, θ, and χ are both the parameters of coordinate rotation and angles of physical rotation. The diatomic molecule is perhaps the most complicated actual system in which a single set of Euler angles can serve both purposes. For example, in polyatomic theory the Euler angles ϕ, θ, and χ are physical rotations describing a frame to which the nuclei are attached, and are not the parameters of coordinate rotation α, β, and γ one would use to demonstrate rotational invariance and conservation of the total angular momentum in a polyatomic model which included vibrational angular momentum.

The Wigner–Eckart theorem breaks the transition moment into two parts, the so-called reduced matrix element whose value is controlled by the initial and final total angular quantum numbers and the degree of transition tensor operator, and Clebsch–Gordan coefficient whose value is controlled by the magnetic quantum numbers and the indices of the components of the tensor operator. The exact separation of the total angular momentum in the Wigner–Witmer eigenfunction simplifies and improves the accuracy of computation of the Hönl–London strengths. Line positions $\tilde{\nu}$ and Hönl–London factors are normally independently evaluated. In the present algorithm, they are simultaneously computed. Of the myriad upper and lower term differences, only those for which the Hönl–London line-strength factor is non-vanishing become spectral lines.

The Born–Oppenheimer approximation is not totally eliminated from computation of a diatomic spectrum. It is needed to break the electronic–vibrational eigenfunction into the product of electronic and vibrational eigenfunction. The Wigner–Witmer diatomic eigenfunction exactly separates the diatomic line strength into the product of the electronic–vibrational strength $S_{n'v',nv}$ and the Hönl–London line-strength factor $S_{J',J}$,

$$S_{n'v'J',nvJ} = S_{n'v',nv}\, S_{J',J}, \tag{H.8}$$

$$S_{n'v',nv} = |a_0 + a_1\bar{r}_{v'v} + a_2\bar{r}^2_{v'v} + \cdots|^2\; q_{v'v}. \tag{H.9}$$

The Born–Oppenheimer approximation separates the electronic–vibrational strength into the product of the square of the electronic transition moment,

$$R_e(r) = a_0 + a_1 r + a_2 r^2 + \cdots, \tag{H.10}$$

evaluated at the r-centroids $\bar{r}^n_{v'v}$,

$$\bar{r}^n_{v'v} = \frac{\langle v'\,|r^n|v\rangle}{\langle v'\,|v\rangle}, \tag{H.11}$$

and with the Franck–Condon factors $q_{v'v}$,

$$q_{v'v} = \langle v'\,|v\rangle^2. \tag{H.12}$$

Whereas experimental and calculated line positions are often known with an accuracy of 1: 10^6 or better, spectral line intensity is rarely recorded with an accuracy of better than 1%. Thus, use of the Born–Oppenheimer approximation for calculation of the electronic–vibrational strength, but not for calculation of line positions, does not introduce significant error.

H.3 Determination of the molecular parameters

Implementation of the algorithm described above requires a comprehensive set of accurate molecular parameters. This is not a minor provison, and the situation is further complicated by the fact that the use of the Wigner–Witmer eigenfunction modifies the manner in which diatomic parameters are determined from accurate line position measurements such as those provided by Fourier transform spectroscopy. The basic idea in both current practice [27, 28] and our determination of diatomic parameters is the calculation of matrix elements of the form

$$\langle a_i\,|H_k|a_j\rangle = \langle n_i v_i J_i M_i \Omega_i \Lambda_i S_i \Sigma_i\,|H_k|n_j v_j J_j M_j \Omega_j \Lambda_j S_j \Sigma_j\rangle \tag{H.13}$$

in which $|a\rangle$ is a Hund's case (a) basis function,

$$\begin{aligned} |a\rangle &= |nvJM\Omega\Lambda S\Sigma\rangle \\ &= \sqrt{\frac{2J+1}{8\pi^2}}\langle\rho, \zeta, \mathbf{r}'_2, \ldots, \mathbf{r}'_N, r\,|nv\rangle\, D^{J^*}_{M\Omega}(\phi, \theta, \chi)\,|S\Sigma\rangle \end{aligned} \quad \text{(H.14)}$$

and H_k is a term from the diatomic Hamiltonian

$$H = \sum_k H_k. \quad \text{(H.15)}$$

Use of the Wigner–Witmer eigenfunction breaks required matrix elements of the Hamiltonian into two parts, one the angular momentum part which can be calculated exactly and the electronic–vibrational part which, except for the very simplest molecules, cannot be calculated with spectroscopic accuracy. Use of a Born–Oppenheimer eigenfunction breaks calculation into three parts, an electronic part, a part consisting of a sum over an infinite number of Born–Oppenheimer vibrational states, and a rotational part. In general, none of the three can be done exactly. Van Vleck transformations, parity partitioning, and the concept of an 'effective Hamiltonian' reduce the dimensions of the Hamiltonian matrix. We simply compute a Hamiltonian matrix composed of case (a) matrix elements, and let the dimension of the matrix determined by the range of Ω in equation (H.5) be that required to make computed line positions $\tilde{nu}_{cal}$ equal the experimental positions $\tilde{\nu}_{exp}$ to within the estimated accuracy of the $\tilde{\nu}_{exp}$. With the exceptions (1) our Hamiltonian of unmodified case (a) matrix elements replaces the effective Hamiltonian used by others and (2) instead of using coded selection rules we use non-vanishing of the Hönl–London line-strength factor as the only selection rule, our determination of molecular parameters is identical to that described by [29]. Trial values of upper and lower parameters are assumed, the line positions are computed, corrections to the parameters are computed from the errors in the computed line positions,

$$\Delta\tilde{\nu} = \tilde{\nu}_{cal} - \tilde{\nu}_{exp}, \quad \text{(H.16)}$$

and the process is repeated until the corrections to the parameters become negligibly small. The algorithm for finding molecular parameters by fitting computed line positions to measured line positions is described in more detail in [30].

H.4 Discussion

The Wigner–Witmer paper [15] appeared about a year after the Born–Oppenheimer paper [16]. The Born and Oppenheimer work treats all molecules, but the Wigner–Witmer paper is strictly limited to the diatomic molecule. Although Born's formulation [16, 31] in terms of

$$\kappa = \left(\frac{m}{M_0}\right)^{1/4} \quad \text{(H.17)}$$

in which m is the electronic mass and 'where M_0 can taken as any one of the nuclear masses or their mean' [31] is rarely used, the Born–Oppenheimer approximation became the foundation of molecular theory. Although the Born–Oppenheimer approximation is applicable to all molecules, in their final section Born and Oppenheimer show that the polar angle and azimuthal angle of the polar coordinates r, θ, and ϕ of the internuclear vector in a diatomic molecule are exactly separable in the spherical harmonic $Y_{\ell m}(\theta, \phi)$. The Wigner complex conjugate of the D-symbol, $D^{j^*}_{m\omega}(\phi, \theta, \chi)$ is the mathematical extension of the spherical harmonics,

$$Y_{\ell m}(\theta, \phi) = \sqrt{\frac{2\ell + 1}{4\pi}}\, D^{\ell^*}_{m0}(\phi, \theta, 0) \tag{H.18}$$

which allows one to deal with the total angular momentum $|jm\rangle$ rather that just the orbital angular momentum $|\ell m\rangle$. The first half of the Wigner–Witmer paper is the logical extension of the Born–Oppenheimer $Y_{\ell m}(\theta, \phi)$ result to include all three Euler angles in diatomic eigenfunction. The two-part Wigner–Witmer paper became famous for its second part which gives correlation rules relating the electronic states of a diatomic molecule to the $L-S$ coupled states of the separated atoms. Oddly, the exact diatomic eigenfunction with which Wigner and Witmer determined their correlation rules has been ignored. About 40 years passed between publication the Wigner–Witmer paper and entry of the Wigner D-function into the literature of diatomic theory. By then, the Wigner–Witmer diatomic eigenfunction had apparently been forgotten.

References

[1] Hornkohl J O, Woods A C and Parigger C G 2014 *J. Phys.: Conf. Ser.* **548** 012033

[2] Singh J P and Thakur S N (ed) 2019 *Laser-Induced Breakdown Spectroscopy* (New York: Elsevier)

[3] Musazzi S and Perini U (ed) 2014 *Laser-Induced Breakdown Spectroscopy* (Heidelberg: Springer)

[4] Cremers D E and Radziemski L J 2006 *Handbook of Laser-Induced Breakdown Spectroscopy* (New York: Wiley)

[5] Noll R 2011 *Laser-Induced Breakdown Spectroscopy* (Heidelberg: Springer)

[6] Singh J P and Thakur S N (ed) 2007 *Laser-Induced Breakdown Spectroscopy* (Amsterdam: Elsevier)

[7] Miziolek A W, Palleschi V and Schechter I 2006 *Laser-Induced Breakdown Spectroscopy (LIBS)—Fundamentals and Applications* (New York: Cambridge University Press)

[8] Parigger C G 2006 Laser-induced breakdown in gases: experiments and simulation *Laser-Induced Breakdown Spectroscopy (LIBS), Fundamentals and Applications* ed A W Miziolek, V Palleschi and I Schechter (New York: Cambridge University Press) ch 4

[9] Parigger C G, Surmick D M, Helstern C M, Gautam G and Bol'shakov A A 2019 Molecular laser-induced breakdown spectrosocpy *Laser-Induced Breakdown Spectroscopy* ed J P Singh and S N Thakur (New York: Elsevier) ch 7

[10] Merten J, Jones M, Hoke S and Allen S A 2014 *J. Phys.: Conf. Ser.* **548** 012042

[11] Parigger C G, Helstern C M and Gautam G 2019 *Atoms* **7** 7030074

[12] Tiwari P K, Rai N K, Kumar R, Parigger C G and Rai A K 2019 *Atoms* **7** 7030071

[13] Surmick D M, Dagel D J and Parigger C G 2019 *Atoms* **7** 86
[14] Parigger C G, EL Sherbini A M and Splinter R 2019 *J. Phys.: Conf. Ser.* **1253** 012001
[15] Wigner E and Witmer E E 1928 *Z. Phys.* **51** 859
Hettema H 2000 *Quantum Chemistry: Classic Scientific Papers* (Singapore: World Scientific), p 287
[16] Born M and Oppenheimer R 1927 *Ann. Phys.* **84** 457
Hettema H 2000 *Quantum Chemistry: Classic Scientific Papers* (Singapore: World Scientific), p 1
[17] Nemes L, Keszler A M, Hornkohl J O and Parigger C G 2005 *Appl. Opt.* **44** 3661
[18] Nemes L, Hornkohl J O and Parigger C G 2005 *Appl. Opt.* **44** 3686
[19] Rose M E 1995 *Elementary Theory of Angular Momentum* (Mineola, NY: Dover)
[20] Edmonds A R 1974 *Angular Momentum in Quantum Mechanics* 2nd edn (Princeton, NJ: Princeton University Press)
[21] Brink D M and Satchler G R 1993 *Angular Momentum* 3rd edn (Oxford: Oxford University Press)
[22] Biedenharn L C and Louck J D 2009 *Angular Momentum in Quantum Physics* (Cambridge: Cambridge University Press)
[23] Zare R N 1988 *Angular Momentum* (New York: Wiley)
[24] Thompson W J 1994 *Angular Momentum* (New York: Wiley)
[25] Varshalovich D A, Moskalev A N and Khersonskii V K 1988 *Quantum Theory of Angular Momentum* (Singapore: World Scientific)
[26] Kovács I 1969 *Rotational Structure in the Spectra of Diatomic Molecules* (New York: Elsevier)
[27] Lefebvre-Brion H and Field R W 2004 *The Spectra and Dynamics of Diatomic Molecules* (New York: Elsevier/Academic)
[28] Brown J M and Carrington A 2003 *Rotational Spectroscopy of Diatomic Molecules* (Cambridge: Cambridge University Press)
[29] Zare R N, Schmeltekopf A L, Harrop W J and Albritton D L 1973 *J. Mol. Spectrosc.* **46** 37
[30] Hornkohl J O, Nemes L and Parigger C G 2009 Spectroscopy, dynamics and molecular theory of carbon plasmas and vapors *Advances in the Understanding of the Most Complex High-Temperature Elemental System* ed L Nemes and S Irle (Singapore: World Scientific) ch 4
[31] Huang K and Born M 1954 *Dynamical Theory of Crystal Lattices* (Oxford: Clarendon)

Appendix I

Program MorseFCF.for

This appendix lists the program 'MorseFCF.for' and the subroutines 'MorseSubs.for'. This program has been utilized in investigations of titanium monoxide (TiO) spectra [1] that were measured following laser-induced breakdown spectroscopy (LIBS) with 1–100 TW cm^{-2} irradiance. Franck–Condon factors and r-centroids were computed and listed for selected TiO transitions [1] using Morse potentials chosen to best fit the low lying vibrational levels. The electronic transition moments were taken from the most recently reported *ab initio* computations for A–X, B–X, and E–X transitions.

I.1 MorseFCF.for

This program computes Franck–Condon factors and first three *r*-centroids for a Morse potentials and sets of molecular parameters.

```
!       February 12, 2012
!       James O. Hornkohl
!       Program to compute Morse Franck-Condon factors
        program morfcf
        use DeclaredStuff
        real (8) :: qs, r1s, r2s, r3s
!           real (8) q, r1, r2, r3
!           dimension q(0: 100, 0: 100), r1(0: 100, 0: 100), r2(0: 100, 0: 100),
           *r3(0: 100, 0: 100)
        real (8) :: weu, wexeu, Beu, muu
        real (8) :: wel, wexel, Bel, mul
        real (8) :: au, betau, reu, gammau, xu, alpha
```

doi:10.1088/978-0-7503-1989-8ch29

```
       real (8) :: al, betal, rel, gammal, xl
       real (8) :: rmin, rmax, rsave, tstu, tstl, dr, r, psi, r0
       real (8) :: lgnrmu (0:100), lgnrml (0:100)
       real (8) :: lgnorm
       real (8) :: alphavu (0:100), alphavl (0:100)
       real (8) :: yu, yl, x, y
       integer *4 :: vmaxu, vmaxl, npts, i, j, k, lun
       integer *4 :: nv, mv
       character *130 :: prnfil, fcffil
       write (*, 1000)
1000   format (15x, 'Franck - Condon␣Factors␣from␣Morse␣Eigenfunctions')
       write (*, 1005)
1005   format (20x, 'Version␣for␣Microsoft␣32 - bit␣Compiler', /)
       write (*, 1010)
1010   format ('␣␣␣␣␣␣␣␣␣␣␣␣␣␣␣Print␣file␣name␣=␣', $)
       read (*, '(a)') prnfil
       write (*, 1016)
1016   format ('␣␣␣␣␣␣␣␣␣␣␣␣␣␣␣FCF␣file␣name␣=␣', $)
       read (*, '(a)') fcffil
!         write (*, 1020)
!  1020   format (5x, 'Enter constants for upper electronic state. ')
!    20   write (*, 1030)
!  1030   format ('                    we = ', $)
!         read (*, *, err = 20) weu
!    30   write (*, 1040)
!  1040   format ('                  wexe = ', $)
!         read (*, *, err = 30) wexeu
!    40   write (*, 1050)
!  1050   format ('                    Be = ', $)
!         read (*, *, err = 40) Beu
!    50   write (*, 1060)
!  1060   format ('          reduced mass = ', $)
!         read (*, *, err = 50) muu
!    60   write (*, 1070)
!  1070   format ('max vibrational quantum number = ', $)
!         read (*, *, err = 60) vmaxu
       weu = 373.152d0
       wexeu = 1.5112d0
       Beu = 0.1542393d0
       muu = 18.64966011d0
       vmaxu = 9
```

```
      write (*, 1075) weu, wexeu, Beu, muu
1075  Format (4(1pg20.13))
      call const (au, alphavu, lgnrmu, betau, weu, wexeu, Beu, reu, muu,
      *gammau, vmaxu)
!       write (*, 1080)
! 1080  format (5x, 'Enter constants for upper electronic state.')
!  100  write (*, 1030)
!       read (*, *, err = 100) wel
!  110  write (*, 1040)
!       read (*, *, err = 110) wexel
!  120  write (*, 1050)
!       read (*, *, err = 120) Bel
!  130  write (*, 1060)
!       read (*, *, err = 130) mul !
!  140  write (*, 1070)
!       read (*, *, err = 140) vmaxl
      wel = 370.201d0
      wexel = 1.3732d0
      Bel = 0.152233d0
      mul = muu
      vmaxl = 9
      call const (al, alphavl, lgnrml, betal, wel, wexel, Bel, rel, mul,
      *gammal, vmaxl)
 150  write (*, 1170)
1170  format ('          r - min = ', $)
      read (*, *, err = 150) rmin
      if (rmin. eq.0.0) goto 160
      xu = au*dexp (-betau*(rmin - reu))
      alpha = alphavu(vmaxu)
      lgnorm = lgnrmu(vmaxu)
      tstu = psi (vmaxu, xu, alpha, lgnorm)
      xl = al*dexp (-betal*(rmin - rel))
      alpha = alphavl(vmaxl)
      lgnorm = lgnrml(vmaxl)
      tstl = psi (vmaxl, xl, alpha, lgnorm)
      write (*, 1180) tstu, tstl
1180  format (' At trial value of r the eigenfunction = ', 1pg11.4, 3x, 1pg11.4)
      rsave = rmin
      goto 150
 160  rmin = rsave
 170  write (*, 1190)
1190  format ('          r - max = ', $)
```

```
      read (*, *, err = 170) rmax
      if (rmax. eq.0.0) goto 180
      xu = au * dexp (-betau*(rmax - reu))
      alpha = alphavu(vmaxu)
      lgnorm = lgnrmu(vmaxu)
      tstu = psi (vmaxu, xu, alpha, lgnorm)
      xl = al* dexp (-betal*(rmax - rel))
      alpha = alphavl(vmaxl)
      lgnorm = lgnrml(vmaxl)
      tstl = psi (vmaxl, xl, alpha, lgnorm)
      write (*, 1180) tstu, tstl
      rsave = rmax
      goto 170
180   rmax = rsave
190   write (*, 1200)
1200  format (' increment in r = ', $)
      read (*, *, err = 190) dr
      npts = idnint ((rmax - rmin)/dr) + 1
      if (prnfil. ne . ' ') then
        open (10, file = prnfil, status = 'unknown')
        lun = 10
      else
        lun = 0
      endif
      write (lun, 1000)
      write (lun, 1090) reu, rel
1090  format (6x, '            re = ', 1pg17.10, 3x, 1pg17.10)
      write (lun, 1100) betau, betal
1100  format (6x, '          beta = ', 1pg17.10, 3x, 1pg17.10)
      write (lun, 1110) gammau, gammal
1110  format (6x, '          gamma  = ', 1pg17.10, 3x, 1pg17.10)
      write (lun, 1120) weu, wel
1120  format (6x, '            we = ', 1pg17.10, 3x, 1pg17.10)
      write (lun, 1130) wexeu, wexel
1130  format (6x, '          wexe = ', 1pg17.10, 3x, 1pg17.10)
      write (lun, 1140) Beu, Bel
1140  format (6x, '            Be = ', 1pg17.10, 3x, 1pg17.10)
      write (lun, 1150) muu, mul
1150  format (6x, '   reduced mass = ', 1pg17.10, 3x, 1pg17.10)
      write (lun, 1210) rmin, rmax
1210  format (6x, '      r - min = ', 1pg11.4, 2x, 'r - max = ', 1pg11.4)
      write (lun, 1220) dr
```

```
1220     format (6x, '␣␣␣␣␣␣delta  r␣ = '1pg11.4)
          write (lun, 1230) npts
1230     format (6x, 'number␣of␣points␣ = ', i4)
          allocate (q(0: vmaxu, 0: vmaxl), r1(0: vmaxu, 0: vmaxl),
          *r2(0: vmaxu, 0: vmaxl), r3(0: vmaxu, 0: vmaxl))
          r0 = dble (idint (50.0d0*(reu + rel)))/100.0d0
          do 220 i = 0, vmaxu
           do 210 j = 0, vmaxl
              qs = 0d0
              r1s = 0d0
              r2s = 0d0
              r3s = 0d0
              do k = 1, npts
                 r = rmin + (k - 1)*dr
                 xu = au*dexp (-betau*(r - reu))
                 alpha = alphavu(i)
                 lgnorm = lgnrmu(i)
                 yu = psi (i, xu, alpha, lgnorm)
                 xl = al*dexp (-betal*(r - rel))
                 alpha = alphavl(j)
                 lgnorm = lgnrml(j)
                 yl = psi (j, xl, alpha, lgnorm)
                 y = yu * yl
                 qs = qs + y
                 x = r - r0
                 y = y*x
                 r1s = r1s + y
                 y = y*x
                 r2s = r2s + y
                 y = y*x
                 r3s = r3s + y
              end do
              qs = qs*dr
              r1(i, j) = r1s*dr/qs
              r2(i, j) = r2s*dr/qs
              r3(i, j) = r3s *dr/qs
              q(i, j) = qs*qs
              nv = i - 1
              mv = j - 1
 210       continue
 220      continue
          call prnfcf (vmaxu, vmaxl, r0, lun)
```

```
          if (lun. ne.0) close (lun)
!        De = weu*weu / (4d0*wexeu)
          if (fcffil. ne . '␣ ') then
             open (unit = 11, file = fcffil)
             do i = 0, vmaxu
                do j = 0, vmaxl
                   write (11, 2000) i, j, q(i, j), r1(i, j), r2(i, j), r3(i, j)
                end do
             end do
             close (11)
          endif
2000     format (1x, i3, ", ", i3, 4(", ", 1x, 1pg14.7))
         end
         subroutine prnfcf (vmaxu, vmaxl, r0, lun)
         use DeclaredStuff
         implicit none
         integer *4vmaxu, vmaxl, lun
!          real (8) q, r1, r2, r3
!          common / block1 / q(0: 100, 0: 100), r1(0: 100, 0: 100), r2(0: 100, 0: 100),
!          cr 3 (0: 100, 0: 100)
         real (8) r0, sum
         integer *4i, j, jmin, jmax
         write (lun, 1000) r0
1000     format (/, 6x, 'r0␣ = ', 1pg12.5 ,/)
         jmin = 0
         jmax = 4
  10     if (jmax. gt . vmaxl) jmax = vmaxl
         write (lun, 1010) (j, j = jmin, jmax)
         do 30 i = 0, vmaxu
          write (lun, 1020) i, (q(i, j), j = jmin, jmax)
          write (lun, 1030) (r1(i, j), j = jmin, jmax)
          write (lun, 1030) (r2(i, j), j = jmin, jmax)
          write (lun, 1030) (r3(i, j), j = jmin, jmax)
  30     continue
         if (jmax. eq . vmaxl) goto 40
         jmin = jmin + 5
         jmax = jmax + 5
         goto 10
1010     format (/, 5x, 'v' '␣', 4x, 'v"␣ = ', i3, 7x, 'v"␣ = ', i3, 7x, &&'v"␣ = ',
         *i3, 7x, 'v"␣ = ', i3, 7x, 'v"␣ = ', i3)
1020     format (2x, i4, 5(3x, 1pg11.4))
1030     format (6x, 5(3x, 1pg11.4))
  40     write (lun, 1040)
1040     format (/, 6x, 'Emission␣sums')
         do 60 i = 0, vmaxu
```

```
       sum = 0d0
       do 50 j = 0, vmaxl
         sum = sum + q(i, j)
50     continue
       write (lun, 1050) i, sum
60    continue
1050  format (6x, i5, 2x, 1pg22.15)
      write (lun, 1060)
1060  format (/,6x, 'Absorption␣sums')
      do 80 j = 0, vmaxl
         sum = 0d0
         do 70 i = 0, vmaxu
            sum = sum + q(i, j)
70       continue
         write (lun, 1050) j, sum
80    continue
      return
      end
```

I.2 MorseSubs.for

This section list subroutines for the main program MorseFCF that computes Franck–Condon factors and the first three r-centroids.

```
       subroutine const (a, b, lgnorm, beta, we, wexe, Be, re, mu, gamma,
       c vmax)
!      July 8, 2011
!      James O. Hornkohl
!      Compute some constants needed for Morse eigenfunctions.
      implicit real *8 (a - h, o - z)
      real *8 h, c, Avog, pi, hbar, x
      real *8 a, b, lgnorm, beta, we, wexe, be, re, mu, gamma
      integer k, v, vmax
      dimension b (0: 100), lgnorm (0: 100)
      real *8 dlgama
      h = 6.6260693d - 27
      c = 2.99792458d + 10
      Avog = 6.0221415d + 23
      pi = dacos (-1d0)
      hbar = h/(2d0*pi)
      beta = 1d - 08*sqrt (4d0*pi*c*(mu/Avog)*wexe
     c/hbar)
```

```
!       Compute equilibrium internuclear distance
        gamma = 1d + 16*hbar*Avog/(4d0*pi*c*mu)
        if (Be. ne.0d0) then
            re = sqrt (gamma/Be)
        else
            Be = gamma/(re*re)
        end if
        a = we/wexe
!       Compute logarithm of normalization factor.
        do v = 0, vmax
!       b(v) is the constant alpha in the Laguerre polynomial.
           b(v) = a - 2d0*v - 1d0
!       lgnorm(v) is the logarithm of the normalization
        *factor for the
!       Morse eigenfunctions.
           lgnorm(v) = dlog(beta)
           if (v. lt.2) goto 10
           do k = 2, v
             lgnorm(v) = lgnorm(v) + dlog (1d0*k)
           enddo
 10     lgnorm(v) = lgnorm(v) + dlog (b(v))
        x = b(v) + v + 1d0
        lgnorm(v) = lgnorm(v) - dlgama(x)
        lgnorm(v) = lgnorm(v)/2d0
    enddo
    return
    end

    real *8 function psi (v, x, alpha, lgnorm)
!    December 28, 2011
!    Compute eigenfunction for Morse oscillator. Subroutine const
!    must be called once before calls to here are made.
!    The vibrational quantum number is n - 1.
      implicit none
      real *8x, alpha, lgnorm, laguer
      integer k, kmax, v
      psi = dexp (lgnorm + alpha*dlog(x)/2d0 - x/2d0)*laguer (v, x, alpha)
      return
      end
```

```
      real *8 function laguer(n, x, alpha)
!     December 28, 2011
!     The degree of the Laguerre polynomial is n - 1.
      implicit none
      real * 8 x, alpha
      integer n, k, kmax
      real * 8 poly(0: 100)
      poly(0) = 1d0
      poly(1) = alpha + 1d0 - x
      if (n. lt.2) goto 20
      kmax = n - 1
      do 10 k = 1, kmax
        poly(k + 1) = ((2*k + alpha + 1d0 - x)*poly(k) - (k + alpha)*
     c   poly(k - 1))/(k + 1)
10     continue
20     laguer = poly(n)
      return
      end
      real *8 FUNCTION DLGAMA(X)
!                        SPECIFICATIONS FOR ARGUMENTS
      real *8  X
!                        SPECIFICATIONS FOR LOCAL VARIABLES
      real *8  P1(9), Q1(8), P2(9), Q2(8), P3(9), Q3(8), P4(7)
      INTEGER       IER, J
      real *8  BIG1, XINF, PI,
      *                   Y, T, R, SIGN, A, B, TOP, DEN, EPS
      LOGICAL            MFLAG
!                        COEFFICIENTS FOR MINIMAX
!                        APPROXIMATION TO LN(GAMMA(X)),
!                        0.5. LE. X. LE.1.5
      DATA          P1(1)/6.304933722864032D02/,
      *             P1(2)/1.389482659233250D02/,
      *             P1(3)/ - 2.331861065739548D03/,
      *             P1(4)/ - 2.651470392943388D03/,
      *             P1(5)/ - 8.953073589022869D02/,
      *             P1(6)/ - 9.229503102917111D01/,
      *             P1(7)/ - 1.940352203312667D00/,
      *             P1(8)/4.368019694395194D00/,
      *             P1(9)/1.279153645893113D02/
      DATA          Q1(1)/6.689575153359349D02/,
      *             Q1(2)/2.419887329355996D03/,
      *             Q1(3)/3.354196974608081D03/,
      *             Q1(4)/1.860416170944268D03/,
      *             Q1(5)/3.944307810159532D02/,
      *             Q1(6)/2.682132440551618D01/,
```

```
     *          Q1(7)/3.440812622259858D - 01 /,
     *          Q1(8)/5.948212550303777D01/
!                 COEFFICIENTS FOR MINIMAX
!                 APPROXIMATION TO LN(GAMMA(X)),
!                 1.5. LE. X. LE.4.0
     DATA       P2(1)/1.071722590306920D04 /,
     *          P2(2)/6.527047912184606D04 /,
     *          P2(3)/1.176398389569621D05 /,
     *          P2(4)/ - 9.726314581896472D03 /,
     *          P2(5)/ - 1.266094622188023D05 /,
     *          P2(6)/ - 5.393468741199669D04 /,
     *          P2(7)/ - 3.895916159676326D03 /,
     *          P2(8)/5.397392180667399D00 /,
     *          P2(9)/5.334390026324024D02/
     DATA       Q2(1)/5.314589562326176D03 /,
     *          Q2(2)/5.493654949398033D04 /,
     *          Q2(3)/2.205757574602192D05 /,
     *          Q2(4)/3.602313576600391D05 /,
     *          Q2(5)/2.273446951911101D05 /,
     *          Q2(6)/4.560612434396495D04 /,
     *          Q2(7)/1.702062439974796D03 /,
     *          Q2(8)/1.670328399370593D02/
!                 COEFFICIENTS FOR MINIMAX
!                 APPROXIMATION TO LN(GAMMA(X)),
!                 4.0. LE. X. LE.12.0
     DATA       P3(1)/ - 6.114039864945718D07 /,
     *          P3(2)/ - 5.588132821261888D08 /,
     *          P3(3)/ - 9.078970022444525D08 /,
     *          P3(4)/3.662935130796460D09 /,
     *          P3(5)/4.658700336821218D09 /,
     *          P3(6)/ - 4.570725249206307D09 /,
     *          P3(7)/ - 2.220833171087439D09 /,
     *          P3(8)/ - 1.474322990113017D04 /,
     *          P3(9)/ - 1.959850795570400D06/
     DATA       Q3(1)/ - 5.636057205056241D05 /,
     *          Q3(2)/ - 2.691827587118628D07 /,
     *          Q3(3)/ - 4.411606716771217D08 /,
     *          Q3(4)/ - 2.774890551941383D09 /,
     *          Q3(5)/ - 6.579874397740792D09 /,
     *          Q3(6)/ - 4.980644951174248D09 /,
     *          Q3(7)/ - 6.677373781427094D08 /,
     *          Q3(8)/ - 2.722530175870899D03/
```

```
!                   COEFFICIENTS FOR MINIMAX
!                   APPROXIMATION TO LN(GAMMA(X)),
!                   12.0. LE. X
      DATA          P4(1)/8.40596949829D-04/,
      1             P4(2)/-5.9523334141881D-04/,
      2             P4(3)/7.9365078409227D-04/,
      *             P4(4)/-2.777777777769526D-03/,
      *             P4(5)/8.333333333333333D-02/,
      *             P4(6)/9.189385332046727D-01/,
      6             P4(7)/-1.7816839846D-03/
      DATA          XINF/.4494232D+308/
      DATA          EPS/.2220446050D-15/
      DATA          PI/3.141592653589793D0/
      DATA          BIG1/1.28118D305/
!                   FIRST EXECUTABLE STATEMENT
      IER = 0
      MFLAG = .FALSE.
      T = X
      IF (DABS(T). LT. BIG1) GOTO 5
      IER = 129
      DLGAMA = XINF
      GOTO 9000
5     IF (T. GT.0.0D0) GOTO 20
      IF (T. LT.0.0D0) GOTO 10
      IER = 130
      DLGAMA = XINF
      GOTO 9000
!                   ARGUMENT IS NEGATIVE
10    MFLAG = .TRUE.
      T = -T
      R = DINT(T)
      SIGN = 1.0D0
      IF (DMOD (T, 2.0D0). EQ.0.0D0) SIGN = -1.0D0
      R = T - R
      IF (R. NE.0.0D0) GOTO 15
      IER = 130
      DLGAMA = XINF
      GOTO 9000
!                   ARGUMENT IS NOT A NEGATIVE INTEGER
15    R = PI/DSIN(R*PI)*SIGN
      T = T + 1.0D0
      R = DLOG (DABS(R))
!                   EVALUATE APPROXIMATION FOR
!                   LN(GAMMA(T)), T. GT.0.0
20    IF (T. GT.12.0D0) GOTO 60
      IF (T. GT.4.0D0) GOTO 50
```

```
      IF (T. GE.1.5D0) GOTO 40
      IF (T. GE.0.5D0) GOTO 25
!                             0.0. LT. T. LT.0.5
      B = -DLOG(T)
      A = T
      T = T + 1.0D0
      IF (A. GE . EPS) GOTO 30
      DLGAMA = B
      GOTO 9005
!                             0.5. LE. T. LT.1.5
 25    TOP = T - 0.5D0
      B = 0.0D0
      A = TOP - 0.5D0
 30    TOP = P1(8)*T + P1(9)
      DEN = T + Q1(8)
      DO 35J = 1, 7
         TOP = TOP*T + P1(J)
         DEN = DEN*T + Q1(J)
 35    CONTINUE
      Y = (TOP/DEN)*A + B
      IF (MFLAG)Y = R - Y
      DLGAMA = Y
      GOTO 9005
!                             1.5. LE. T. LE.4.0
 40    B = T - 1.0D0
      TOP = P2(8)*T + P2(9)
      DEN = T + Q2(8)
      A = B - 1.0D0
      DO 45J = 1, 7
         TOP = TOP*T + P2(J)
         DEN = DEN*T + Q2(J)
 45    CONTINUE
      Y = (TOP/DEN)*A
      IF (MFLAG)Y = R - Y
      DLGAMA = Y
      GOTO 9005
!                             4.0. LT. T. LE.12.0
 50    TOP = P3(8)*T + P3(9)
      DEN = T + Q3(8)
      DO 55J = 1, 7
      TOP = TOP*T + P3(J)
      DEN = DEN*T + Q3(J)
```

```
55    CONTINUE
      Y = TOP/DEN
      IF (MFLAG)Y = R - Y
      DLGAMA = Y
      GOTO 9005
!                     12.0. LT. X. LT. BIG1
60    TOP = DLOG(T)
      TOP = T*(TOP - 1.0D0) - .5D0*TOP
      T = 1.0D0/T
      Y = TOP
      IF (T. LT . EPS) GOTO 70
      B = T*T
      A = P4(7)
      DO 65J = 1, 5
         A = A*B + P4(J)
65    CONTINUE
      Y = A*T + P4(6) + TOP
70    IF (MFLAG)Y = R - Y
      DLGAMA = Y
      GOTO 9005
9000  CONTINUE
      write (*, '(&␣Error␣in␣DLGAMA&)')
9005  RETURN
      END
```

Reference

[1] Parigger C G, Woods A C, Keszler A, Nemes L and Hornkohl J O 2012 *AIP Conf. Proc.* **1464** 628

Lightning Source UK Ltd.
Milton Keynes UK
UKHW050626070421
381422UK00006B/91

9 780750 318907